PRENTICE-HALL, INC., *Englewood Cliffs, New Jersey*

CONSTRUCTION: A Guide for the Profession

THOMAS A. GROW
Department of Civil Engineering
University of Massachusetts

Library of Congress Cataloging in Publication Data

Grow, Thomas A
Construction: a guide for the profession.

Bibliography: p.
1. Contractors' operations. I. Title.
TA210.G7 624 74-23806
ISBN 0-13-169326-3

10 9 8 7 6 5 4 3

Printed in the United States of America

PRENTICE-HALL INTERNATIONAL, INC., *London*
PRENTICE-HALL OF AUSTRALIA, PTY. LTD., *Sydney*
PRENTICE-HALL OF CANADA, LTD., *Toronto*
PRENTICE-HALL OF INDIA PRIVATE LIMITED, *New Delhi*
PRENTICE-HALL OF JAPAN, INC., *Tokyo*

Dedicated to

MY WIFE
AND CHILDREN

Contents

CHAPTER 6
Specifications

CHAPTER 7
Estimating

CHAPTER 8 Quantity Survey-Small Building 113

CHAPTER 9 Earthmoving 137

Preface

This book is partly the result of suggestions made by students in a course taught by the author to Civil Engineering majors at the University of Massachusetts for the past fifteen years.

Construction is the end product of civil engineering design, yet many students and beginning engineers have only a vague idea of the work and exercise of talent that must be expended in transforming their calculations and design drawings into highways, dams, buildings, airports, or water and sewage treatment facilities. A proper appreciation of the part played by each member of the construction team, consisting of owner, architect or engineer, and contractor, is necessary for the efficient functioning of the team in performing engineering construction; yet too many members of the team are not aware of the problems facing the others. The lack of tolerance and the ignorance of the responsibilities and duties displayed by members of the team toward each other are the causes of many unsatisfactory construction projects and much costly and wasteful litigation.

At a time when contractors are hiring more engineers than ever before, the engineering school curricula are becoming more and more mathematical and scientific with less time available for studies related to the construction industry. Fortunately, the number of colleges, universities, and technical schools offering courses in Construction Engineering is increasing and these schools offer many excellent courses pertaining to the work of the construction contractors.

PREFACE

This book is intended to serve those who are not definitely committed to a comprehensive study of the construction industry. It is hoped that the material covered will be of interest to students in technical institutes as well as colleges and universities. The number of topics introduced is rather large in order to provide a survey of some of the varied problems facing members of the construction industry. No topic is covered completely or exhaustively in order that an instructor may expand on some topics and hold coverage of others to a minimum. The meager treatment of some of the topics can be appreciated if we note that an entire college course can be devoted to estimating, another to specification writing, another to business law, etc.

It is hoped that the problems discussed are important to all members of the construction team. No effort has been made to favor the view point of one over the other two.

Much of the book is descriptive in nature. The chapters on estimating, the critical path method, and bidding are more quantitative and contain numerical problems that require the use of simple arithmetic. The only technical material is the short chapter on concrete formwork and even that requires only sophomore-level knowledge of strength of materials.

THOMAS A. GROW

CHAPTER 1

The Construction Industry

1.1

ECONOMIC IMPORTANCE

The construction industry has become this country's largest industry, directly or indirectly furnishing employment to more than 15 per cent of the total working force of the nation. In dollar value even agriculture is now exceeded by the value of construction. In 1969 the value of new construction was $91 billion; the gross national product that year was about $920 billion. By 1973 new construction had increased to $136 billion, about 10.5 per cent of the $1,304 billion gross national product. Repairing, remodeling, and maintenance will add about one third to these figures, so the total construction for 1973 was about $175 billion. New construction for 1974 is expected to increase to about 11 per cent of the gross national product. The value of future construction is sensitive to inflation, fuel shortages, and changes in world economic conditions. Uncertainty concerning the future will lead to an increase in the amounts planned for repairing and remodeling and a corresponding decrease in new construction, resulting in a lower total volume.

1.2 TYPES OF CONSTRUCTION

Construction may be classified in many ways, but one of the most popular methods classifies all construction as one of the following three types:

Buildings educational, commercial, light industrial, social, housing, recreational, hospitals.

Highways bridges and other appurtenances, signs, excavating, filling, grading, paving, seeding and planting, fences.

Heavy construction dams, water and sewage works, airports, docks and other waterfront structures, tunnels, power plants, industrial plants.

Since nearly one third of all construction is paid for by public funds, construction is also often classified as either public or private work. In private work the purpose of the construction is usually to make a profit and the method of contracting the work can often be rather casual. In public works, however, a need or demand rather than the hope for profit is the motive, and public policy places many restrictions on the contracting method. Both buildings and heavy construction may be either public or private, but most highway work must be classified as public works.

1.3 CONSTRUCTION COMPANIES

Construction companies range in size from the small company that builds a few one-family homes to the giants that annually do more than $1 billion worth of work. There may be only one or two employees or, in the large firms, there may be hundreds of permanent workers representing the various building trades as well as managers, engineers, clerks, accountants, architects, attorneys, and planners.

The trend at present is for construction companies to become even larger. We have not yet reached the point, as in the automotive industry, where three or four companies dominate the field, but each year the largest continue to become even larger. In 1968 the largest contractors did $28.6 billion worth of work, an increase of 17 per cent over the worth of the top 400 during 1967. The gains made by the top 400 since then have shown a steady increase. *Engineering News-Record* in its April 11, 1974 issue reports that the top 400 did $55 billion of work in 1973, a 38 per cent increase over the $40 billion reported in 1972.

1.4 COMPANY ORGANIZATION

The simplest type of company organization is the single proprietorship, in which one person controls the business. The company can be formed or discontinued as the owner sees fit. The proprietorship offers the advantages of freedom from many government restrictions, complete freedom of action, tax savings under certain conditions, and complete control. Disadvantages include unlimited liability, often restricted capital, and high taxes under some conditions.

A partnership results when two or more individuals combine their talents and resources into a business enterprise and share in the profits or losses. A written agreement defining the duties and responsibilities of each partner is usually necessary to ensure efficient operation of the business.

The most common type of partnership is one in which each partner has equal responsibility for controlling the business and shares equally in the profits and losses. A secret partner is one who is legally a partner but for some reason conceals his partnership from the public. A silent partner takes no active part in the business but has invested in the firm and shares in the profits and losses. As in a proprietorship, all partners have unlimited liability.

Exceptions to this rule are special or limited partners whose liabilities are limited by the amount invested and who seldom take an active role in management. Special partnerships are restricted by the laws of the various states.

A partnership is not a legal entity and cannot own real estate and does not pay income taxes. Instead, each partner pays taxes on his own share of the profits. Title to real estate, although the property may be used for business purposes, remains in the name of one of the partners. Any partner may make agreements for the company within the scope of the business, and each partner is liable for all the debts of the company. A creditor may take all the property of the business and, if necessary, may take personal property of one or all of the partners. The greatest advantage of a partnership over a proprietorship comes from the larger scope of business possible because of combining financial and personal assets of two or more parties. The unlimited liability of each partner for all the company's debts is a real disadvantage in case one partner proves to be incompetent or dishonest.

The problems arising when one partner dies or otherwise becomes incapable of activity often terminate a successful business. Since a partnership is an agreement between two or more parties to do business as one company, the death of one partner automatically terminates the partnership. For the company to continue in business, the surviving partners may form a new partnership or may acquire new partners, but in any case the value of the shares of ownership held by the deceased become the property of his heirs. Partners often insure the life of other partners to make certain that the survivors can continue in business upon the death of one.

The duration of a partnership may be defined in the partnership agreement. When no definite duration has been established, the partnership may be terminated by mutual agreement, notice by one partner to the other, bankruptcy, or by court decree. The partnership agreement may be changed at any time by the mutual consent of all the partners.

A corporation is an artificial being created by law for the purpose defined by the corporate charter. Corporations, although composed of one or more parties, have special or corporate names and are regarded as separate individuals in all business dealings.

The corporation can own real estate, enter into contracts, incur debts, be sued, sue others—all in its own name and upon its own responsibility. The type of business it may engage in and other duties, responsibilities, and limitations are defined by its corporate charter.

The owners of a corporation are the stockholders who have purchased shares of stock and are entitled to shares of the profits of the company and the possibility of losing their investments if the business fails. The liability of a stockholder, unlike a member of a partnership, is limited by his investment; that is, he cannot lose more than he paid for his stock. This is one of the great advantages of the corporate over other forms of organization. Stock can be bought and sold without the consent of other stockholders or the corporation itself. The corporation is thus of indefinite duration, and the death of its stockholders has no effect upon the life of the corporation.

The management of a corporation is performed by a board of directors elected by the stockholders, usually at an annual meeting where stockholders have votes conforming to the number of shares of stock owned. The board of directors is responsible for the general policy-making decisions required in the business, but the detailed management of the company is in the hands of the officers, such as president, treasurer, secretary, and vice presidents, who are appointed by the directors.

Advantages of corporate organization are its perpetual existence, the limited liability of stockholders, the relative ease of raising capital by the issue and sale of additional stock when necessary. The limited scope of the business as defined by the charter can be a disadvantage. The lack of direct control over management by the individual stockholder can be considered either an advantage or disadvantage, depending upon circumstances. The double taxation imposed upon corporations is a disadvantage since the corporation, as an artificial being, pays income tax on its profits, and the remaining profits, after being distributed to stockholders in the form of dividends, are once again taxed as part of the stockholders' income.

Construction corporations are private corporations since they are composed of private citizens who have invested in the corporation for the purpose of making a profit. Public corporations, on the other hand, are formed by and completely owned by a unit of government. These may exist for building highways, operating schools, building and operating airports, or engaging in any public works project. The duties and responsibilities of the public as well as the private corporations are clearly defined in the corporate charter with the public interest in mind. Water and power companies and other public utilities, if privately owned and operated, are private corporations even though they exist to serve the public interest. They are subject to much more control by government than the usual private corporation since they affect the public interest so greatly.

1.5 COSTS AND COST TRENDS

In a time of inflation, when the cost of wages and services and everything we purchase is increasing, it would seem strange if the cost of construction were to remain constant, and of course it has not. Indeed, the construction industry has been in the lead in the wage-price spiral. At the present time the cost of construction is increasing at a rate of about 10 per cent each year.

Engineering News-Record in its quarterly roundup of construction costs uses its 20-cities Construction Cost Index and its Building Cost Index to show the trend in the past and try to get an idea of the future. The Construction Cost Index uses the cost of common labor and the price of structural steel, portland cement, and lumber 2 x 4's; in computing the Building Cost Index, skilled building trade costs plus the same materials are used. With costs in 1913 considered as 100, these indexes had risen to 1,942 and 1,161, respectively, by December 1973, and as yet there are few signs of any significant leveling off.

The cost of construction is reflected in the "Four M's"—Manpower, Materials, Machinery, and Money. Deferred wage hikes combined with contract settlements have increased skilled wages by more than 50 per cent in the past eight years. Unskilled labor's cost has nearly doubled in the same period. The costs of materials and machinery, although not increasing spectacularly, have increased about 30 per cent in these eight years. The cost of borrowing money has recently reached an all-time high, with 11 per cent and more as the prime rate of interest.

Most construction projects are of somewhat long duration, with the larger ones taking several years to complete. The contractor, in preparing cost estimates for a project, must be able to predict the probable costs of labor and materials a year or more in the future. The cost of construction today is at least 6 per cent higher than a year ago. Estimators today are anticipating a similar rise during the coming year, and their bid prices must include this expected increase for jobs that will be completed during the next year.

In the past the cost of labor has been the main factor in increasing construction costs. With the strong union organization of the building trades there seems to be little prospect of reversing this trend unless, of course, the nation's economy suffers a real setback. One hope for slowing the rise in costs lies in making labor more productive. This can be accomplished by greater use of machinery, changes in building components, use of different materials, and a greater use of factory-fabricated modules.

1.6 CONSTRUCTION-BUSINESS FAILURES

The manner in which contractors get much of their work, by bidding against others, makes the construction industry fiercely competitive, resulting in many contractors working for very low, sometimes inadequate profit margins. The competition and resulting low prices are often cited as a great advantage to the owner of a project. As long as adequate supervision and inspection can ensure the quality of work, as long as plans and specifications are complete and adequate, and as long as the contractor remains financially solvent, this is probably true.

Since the end of World War II, however, an increasingly large number of contractors has gone out of business, owing millions of dollars to creditors. In 1950, for example, 838 companies failed, leaving liabilities of over $27 million. By 1962 failures were up to 2,700. Since 1966 the trend has been toward a lessening of this figure such that for 1970 only 1,687 failures were reported, and 1,419 in 1973. Although the trend seems encouraging, the 1973 failures resulted in a loss of more than $309 million to creditors of the bankrupt companies.

Dun and Bradstreet, Inc., whose studies have made the above figures available, has determined that neglect, fraud, disaster, and lack of experience have caused some failures, but the largest single factor contributing to bankruptcy has been poor management. Incompetence, unbalanced experience, and lack of managerial ability have accounted for more than 80 per cent of the failures.

The construction industry is growing up. It uses expensive and sophisticated machinery and equipment, and the successful construction manager must know how and when to apply them. He must know how and where to purchase materials and equipment. He must be able to recruit and retain both office and field personnel. He must maintain and be able to interpret financial reports. He must be equipped to estimate project costs and make intelligent bids. He must understand legal principles and the operation of government agencies. The construction industry promises great financial as well as intangible rewards to the good manager, but it also promises a sure and expensive defeat for those who enter the industry poorly prepared.

CHAPTER 2

Contracts

Most construction work is done by contract, so an understanding of this legal field is very important to those connected with the construction industry. This chapter will be devoted to contracts in general; applications to construction will be reserved for Chapter 3.

People in all walks of life continually make contracts and perform according to the terms of the contracts. Whenever money or anything of value is exchanged for goods or services, a contract has been made. When a person goes to work in the morning, he does so because of his contract with his employer. When he receives his pay at the end of the week, he does so because the employer is performing according to the terms of his contract. The electricity is carried to our homes because of contracts we have made with the power company. When we buy a loaf of bread we make a contract with the grocer. Contracts can be very simple or they may be very long and complicated legal documents, but the principles involved in all are the same. They can be oral or written or seemingly neither. In purchasing a candy bar from a vending machine, the purchaser inserts a coin in a slot and bar of candy is delivered. No words need be spoken or written, but the candy is delivered because the owner of the machine and the purchaser have made a contract.

2.1
ELEMENTS OF A CONTRACT

According to its simplest definition, a contract is an agreement enforceable at law, but not all agreements are contracts. Five elements must be present before an agreement can become a contract. These elements are discussed in the following paragraphs.

2.2
COMPETENT PARTIES

For an agreement to be a contract there must be two or more competent parties. A party to a contract may be one person, it may be several people acting together as in a partnership, or it may be impersonal as in a corporation. In order to be considered competent, a party must have a certain legal standing. The following parties have their abilities to contract limited in some way, hence are considered incompetent:

Infants An infant or minor is one who has not reached his twenty-first birthday, although the age is less in some states. Minors may contract for necessities, but exactly what constitutes anything reasonably required for health, comfort, and education is often difficult to determine and may vary for different individuals. A contract made with a minor for something not a necessity may be voidable, that is, binding upon the other party only. If a minor avoids a contract he must return any goods or money received under the terms of the contract if it is possible to do so. Some states provide that marriage removes a minor's incompetency; in some others a minor may request a court to declare him competent for the purpose of entering into a business venture.

Persons of unsound mind Contracts made by a person of unsound mind are usually valid, since most are made during a period when he is lucid or sane. Contracts by insane persons for other than the necessities are voidable and may be made either valid or void at his option when he returns to sanity. The other party to the contract does not have this option. Contracts created by law in which one party is not sane are valid, and contracts in which the sane party had no knowledge of the other's insanity and the contract has been completely or partially performed are also valid.

Intoxicated persons A contract made by a person who was sufficiently intoxicated that he could not understand the nature of the contract is voidable by the drunken person. Contracts made for necessities are valid. The degree of intoxication determines the legal status of the contract, and this, as well as the degree of insanity, is difficult to determine. Lack of judgment is the reason for the voidability option of minors, persons of unsound mind, and intoxicated persons. Courts often look upon intoxication as a deliberate act as opposed to being young or insane and are less diligent in protecting the rights of the intoxicated person than the others.

Corporations A corporation is limited by its corporate charter as to the type of activity it may engage in, and any agreement for action beyond that scope of activity is "ultra vires" and will not be enforced by a court of law. If an ultra vires agreement is of the "executed" type, with performance of both parties completed at the time the agreement was made, the courts will enforce the agreement and a valid contract will result. A contract calling for action in the future is an "executory" contract and courts will enforce executory ultra vires agreements only if it is impossible to return to the "status quo" or conditions that existed before the agreements were made. The corporate charter states which officer of the corporation, usually the treasurer or president, has authority to enter into contracts on behalf of the corporation. Contracts made by other officers are ultra vires acts. A corporation is liable for harm or damage to another person or his property, even though such damage may be the result of an ultra vires act.

Professionals Engineers and architects engaged in the design of public works, such as highways, bridges, and buildings in which public health or safety are involved, are required to be registered and have a license for such professional practice. An agreement with an unregistered person for such work is invalid, because such a person is deemed, legally, to be incompetent.

Governments Contracts for work for federal or state governments are made with government agencies through contracting officers who represent those agencies. The project and the funds to finance it and the agency created to administer it must be authorized by legislation. A contract beyond the scope of this specific legislation is void even though the contracting officer acted in good faith when he entered into such a contract.

Agents An agent is a person employed by another, called a principal, to represent him in certain business dealings. The agreement between agent and principal may be oral, written, or merely implied, but contracts made by the agent in the principal's behalf are as valid as if made by the principal himself as long as the agent acts within the scope of the agency. Since contracts made by the agent are for the benefit of the principal, the principal must be competent, but the competency of the agent is unimportant. The principal is responsible for all damages to others and for other acts committed by the agent as long as the agent acts within the scope of the agency. Negligence and lack of diligence by the agent which cause harm to others, even though the agent acts within the scope of the agency, are the liability of the agent. Even though an agent may act beyond the limits of his agency, the principal may ratify such agreements and thereafter be bound by them. When the agent acts beyond the scope of his authority or when the agency character of the transaction is not disclosed, the agent may be liable to the other party of the contract. In the performance of his duties the agent is expected to use ordinary care and skill, but perfection is not required. Mistakes made by the agent, unless he was negligent, are the responsibility of the principal. Securing employment by claiming skills that he does not possess, or by failing to use those skills in fulfilling his duties make the agent liable for harm to third parties as well as harm or damage to his principal.

2.3 PROPER SUBJECT MATTER

For the subject matter of a contract to be proper, the first requirement is that it be clearly defined as to the rights and obligations of each party. A contract cannot be enforced at law unless the court can determine with certainty what is expected in performance by each party. Second, the purpose of the contract must not be contrary to statute law, common law, or public policy. If the terms of the contract are such that the performance would violate statute or common law, courts would rule the contract void. If, however, the contract could have been performed within the law and one of the parties violated the law, the contract would still be valid.

Subjects contrary to either law or public policy include the restraint of trade or creation of monopolies, usurious rates of interest, gambling, sale of public office, any action injurious to public health or safety, harm to third persons or their property, breach of trust or confidence, deception, or anything detrimental to the security of a marriage or home and family relationships.

2.4 CONSIDERATION

The third essential of a contract is that there be a lawful and valuable consideration given by both parties. A consideration, often called "quid pro quo," or something for something, is anything of value. Since it must be lawful, the giving or promise to give a quantity of illegal drug cannot be used as a valid consideration. Courts are not interested in equality of consideration unless the inequality is sufficient to indicate fraud. An exception to this is the case where money is used to liquidate a debt. A money debt cannot be liquidated by a lesser amount unless payment is made before the expected time of payment, and in that case the time is deemed valuable and therefore becomes a part of the consideration. A consideration must be something of value or the promise to give something of value in the future. Giving up a legal right now or in the future is a consideration, even though its value is difficult to determine, and it certainly is not tangible. A gift, now or in the past, cannot classify as consideration since there was nothing given by the other party in return. Saving a person from drowning by either a lifeguard on duty or by one who could save the person with no risk to himself is merely the performance of duty and is not a consideration. Capturing a wanted criminal by a policeman on duty is performance of duty and makes him ineligible to collect a reward, whereas capture by the same man while not on duty would constitute a consideration and make it possible for him to collect the reward.

A consideration must be possible. The collapse and destruction of a bridge would make it impossible to paint, and a contract calling for its painting would be void.

In the past, many documents, especially those of great importance, were sealed by attaching a small piece of wax containing an impression to the documents. Under common law the act of sealing became so important that such sealed contracts did not need a consideration to be valid. In some states today the same is true, even though the actual wax seal is not used and the word "seal" is merely written on the contract.

2.5 AGREEMENT

For a valid contract there must be mutual agreement between the parties, often referred to as a "meeting of the minds." The terms may be very definite, resulting in an express contract, or very indefinite, resulting in an implied contract. Leaving a watch at the jeweler's to be repaired is an example of the latter. The implication is that if the watch is repaired, the owner will pay the repair bill. At the time of making the contract neither the owner nor the repairman has any idea of what is wrong with the watch or how much the bill will be, but each party has a definite idea of his own responsibility, hence a valid contract exists.

The usual method of reaching mutual agreement is when an offer by one party is accepted by the other. An offer is the statement by one party of what he will give or do in return for an act or promise by the other party. The offer may be either verbal or written, and the acceptance must be without change or reservation. If no time limit is specified by the offeror, the other party may assume that he has a reasonable time available in which to accept or reject the offer. The nature of the projected contract determines what the courts will deem "reasonable." When both offer and acceptance are by mail, the offer is made when received by the offeree, but acceptance is made when it is mailed, or given to the U. S. Postal Service for delivery. If an offeree is willing to accept an offer if changes are made, he may submit a counteroffer or counterproposal, but no agreement will be reached until the original offeror accepts the counteroffer. An offer may be made to one person only, in which case only he may accept it. An offer to the public may be accepted by anyone, an example being the offer of a reward for the return of a lost article. An offer may be revoked or withdrawn at any time prior to its acceptance and is automatically revoked if the offeror dies before acceptance or if the terms of the agreement become illegal before being accepted.

Situations sometimes arise in which the mutual agreement of the parties can be questioned. A mutual mistake, resulting in an invalid contract, is made when the two parties to a contract are mistaken concerning material facts involved in a transaction, or have in mind something completely different. Two persons agree on the terms of sale of a quantity of grain. If, unknown to either of them, the storage bin containing the grain had burned down, a mutual mistake had been made and the contract is invalid. In the same transaction, if one person had in mind grain stored on one farm, and the other was thinking of the better-quality grain stored in a different place, there is no meeting of the minds and no contract exists. A unilateral mistake is one in which one party is mistaken as to the true value or condition of an article or in some way does not fully understand the implications of his contract.

Generally such contracts are binding since it is felt that a person should not enter into an agreement he did not fully understand. Clerical errors and mistakes in arithmetic can result in unilateral mistakes in which one party is aware of the mistake and tries thereby to take advantage of the other. Courts endeavor to correct clerical errors in accordance with the intent and original terms of the contract.

Contracts made under duress and those made using undue influence raise the question of reality of consent and result in an invalid contract. Duress is the threat of violence or injury to one's person or property. Such threats impair a person's ability to exercise his will and make contracts voidable. Undue influence results from the trust or confidence that has been established by previous association, or by relations between a person and his family, doctor, attorney, business manager, or other. Improper pressure exerted by any of these and resulting in harmful action is undue influence and results in a voidable contract.

The existence of fraud makes a contract voidable at the option of the victim. Four requirements must be met:

1. There must be an intentional misstatement of fact. The statement must pertain to material fact that is not obvious. It must pertain to the past or the present; it cannot be an expression of opinion concerning the future. Concealing or suppressing information that should have been given to the victim, especially when a previous relationship of trust existed between parties, can be considered a misstatement.

2. There must be intent to deceive.

3. The misstatement results in action by the victim. Action taken by a third party who overheard the misstatement is not fraud. No matter how badly the facts were misrepresented, if the intended victim does not act, there is no fraud. If it can be shown that the intended victim would have acted even without the misstatement, there is no fraud.

4. The resulting action is harmful. Injury to the defrauded must result or there is no fraud.

2.6 PROPER FORM

The last essential of a contract pertains to the form it must have to be enforceable. Most contracts are oral, but good business practice dictates that, except for the very trival ones, the terms of a contract be written so that both parties are very sure of what their rights and responsibilities are. Courts, if they are to enforce contracts, must be sure of their terms, and this leads to much difficulty with oral contracts. Because of the difficulty with oral contracts, the English Parliament in 1677 passed "An Act for the Prevention of Frauds and Perjuries." Most of the provisions of this law have been passed in the states of this country and pertain to the form that must be used in certain types of business contracts. According to this statute, the following must be in writing:

1. A special promise by an executor or administrator to be liable for the debts of the deceased. An executor is appointed in a will to carry out the terms of the will, and an administrator is a person appointed by a court to dispose of the assets of a deceased person if he died without a will. In each case the promise to use his own property on behalf of another is rather unusual and should be in writing.

2. A special promise to answer for a debt or default of another. In this case the promisor is acting as surety for another, such as the cosigner of a note for a bank loan. A promise to pay the debt of another is a primary promise and may be oral. A promise to pay the debt of another if he does not or cannot is a secondary promise and must be written to be enforced.

3. Any contract whose terms cannot be performed in one year must be in writing. The time starts the day after the agreement was made, not when performance of the contract begins. The time is held to be the shortest possible, not the probable time. An oral contract for performance that can be completed in less than a year

may be extended so that the entire transaction takes more than a year to complete without being in writing. A contract to do something or to continue to do something "for life" need not be written, since the promisor's life might not be as much as one year, and the shortest possible rather than the most possible time is the ruling factor.

4. A contract for the sale of real property or its lease for more than one year must be in writing. Property is often classified as being either real or personal property. Real property is fixed, immovable, and imperishable, usually consisting of land and those objects permanently attached to it, such as buildings, trees, fences, minerals, standing water, and water beneath the surface. Running water and growing crops are personal property. Any property not classified as real is personal, often called "chattels." An oral lease involving an interest in land may be valid unless the time of the lease is for more than one year.

5. Contracts for the sale of goods in excess of a fixed amount must be in writing. The original amount in the English statute was £10, but the various states have set their own, and often different, dollar amounts. A contract for services need not be in writing, hence if the goods in question do not yet exist, the manufacture of those goods might constitute a service, and only an oral agreement is necessary. If the nature of the goods is such that they would be salable on the open market, the contract must be in writing. If the goods are specially made for a particular purchaser, their nature is more of a service and an oral contract is sufficient. An oral contract, even though according to the statute of frauds it should have been in writing, may still have some value. If one party to the contract has already fulfilled his obligations, the court may enforce performance by the other party by means of a quasi contract, a contract formed by a court to prevent one party from profiting unduly from the action of another. If a transaction for goods of large-enough value that the terms should be written is covered by oral agreement only, the contract is valid if part of the goods have been delivered and accepted, or if part payment for them has been made.

CHAPTER 3

Construction Contracts

Depending upon the nature of the transaction, contracts may be oral agreements, very simple written documents, or very complex documents in which every aspect of the transaction is spelled out in great detail. Most construction is done by contract, and, as is true in other businesses, such contracts vary greatly in length and complexity. Many professional societies and government agencies have done a great deal toward the standardization of construction contracts such that the general form and content are well established for the various types of construction that might arise. This standardization has resulted in economy in the preparation of the contracts and has also resulted in more satisfactory construction, since the industry has become acquainted with the meaning of the many clauses and conditions of the contract. Many uncertainties have been removed since many of the legal implications of the standard contracts have been tested in court.

The type of construction chosen must be the one that will result in the highest quality work at minimum cost to the owner. Designers who have been successful with a particular type are sometimes reluctant to change, even when changed circumstances might dictate a different type of contract.

3.1 CONSTRUCTION BY FORCE ACCOUNT

The simplest method of construction to visualize is the direct employment of labor, often called the day-labor or force-account method. When using this method, the owner of the project employs workmen, purchases the material and equipment required, and supervises the work himself. In industrial plants and institutions where there are crews of men available for small jobs, this method often results in a satisfactory job, but it is seldom used for large or complicated projects. Projects undertaken by this method are extremely flexible and can start before plans are complete and changes can be made easily as the job progresses.

One of the reasons advanced for this type of construction is that money is saved by the owner because payment of the contractor's overhead and profit are avoided, but this saving is often very hard to realize in practice. When a contractor is employed to complete a project, the owner gets the benefit of the contractor's experience in construction. Hiring adequate labor and the important supervisory personnel is difficult when it is known that the job is to be one of short duration. Purchasing or renting equipment for a short job can be very costly and can often be done more cheaply by a contractor that by a project owner. Knowing where to purchase materials and how to get the best prices are part of a contractor's business, and he can usually do better in these matters than a layman or newcomer to the construction industry.

Although the force-account method does not pay overhead charges or profit directly, some of the inefficiencies of the method may result in a higher total cost than would have been the case if a contractor had done the job and had included overhead and profit as part of his price.

3.2 COMPETITIVE-BID CONTRACTS

A competitive-bid contract results when a prospective owner solicits bids from all interested contractors with the idea of awarding the contract to the lowest bidder. The purpose of the competition is to make sure that the project will be completed for the lowest possible cost to the owner. This free competition between suppliers of goods and services is an important aspect of this nation's economy and usually, but not always, leads to low prices. The two types of competitive bids are as follows:

Lump-sum bid The lump-sum type of bid is used for buildings, typically, in which plans and specifications can be prepared in such detail that it is possible to determine exactly the requirements of the work to be undertaken. The quantities of all the materials and the cost of labor required to install the materials for the entire project can be determined by an estimator. The total cost to the contractor can be determined in advance, and his cost plus allowances for overhead and profit constitute the contractor's bid price. This total bid is then compared with the bids of the competing contractors and the contract awarded to the contractor whose bid is the lowest. The total cost to the owner is thus known before construction starts—a distinct advantage over the situation when a force account is used and the total cost may be completely unknown.

Under this type of contract, changes in plans can be very expensive and a source of trouble unless provision for payment is carefully provided for in the contract. Lump-sum contracts are used when unknown factors are a minimum and when plans and specifications are complete before the bidding starts. The contractor hopes to complete the job as quickly as possible to minimize overhead costs, make his profit, and move on to the next job. The owner hopes to get the best possible construction for the price he has agreed to pay. If the plans and specifications are well done so that both contractor and owner know exactly what is to be done, the result can be very satisfactory to both sides. Strict inspection to ensure that all the terms of the contract are being observed by the contractor is often necessary, especially when the contractor's bid was too low to include adequate profit, or in cases when unforeseen contingencies appear to increase the contractor's costs by a large amount.

Unit-price bid A unit-price contract is used when quantities of a relatively few types of activity are unknown. Highways, dams, airports, and often building foundations are built using a contract of this type. The work for a building foundation may consist of excavating, placing concrete, and backfilling—seemingly only three simple tasks—but soil conditions may be quite different from those anticipated by the designer of the building. The quantity of concrete used to support the building, for example, may be considerably different from that anticipated by the original designer, and the contractor should not be expected to gamble on subsurface conditions. The type of digging and the quantity and hardness of rock encountered are also unknown before excavation starts, and the contractor should be paid more for excavating rock than soft clay.

The unit-price contract promises to pay the contractor a given amount for each cubic yard of clay excavated, a different price for each unit of rock, and a fair price for each unit of concrete in place. These unit prices are determined by the contractor and include his cost plus provision for overhead and profit. In preparing his bid, the contractor is furnished a list of the various activities or job types for the project along with estimated quantities. To get the total cost, each estimated quantity is multiplied by the contractor's unit price, and these extensions are then added. Bids from competing contractors are compared and the lowest total determines the winning bidder; the contract can then be awarded to him.

In the actual construction the contractor is paid according to the quantity of work that he does, not the owner's estimated quantity that was used in bidding. The owner is thus not sure at the outset of the exact total cost of the project. In the unit-price contract, many of the contractor's uncertainties are removed, and he is not required to gamble as much as he would on a lump-sum contract for the same project. This results in lower reserves for contingencies on the part of the contractor and a corresponding lower cost to the owner. Construction can sometimes start before plans are complete since the contractor knows he will be reimbursed at a known rate for each unit completed. Changes involving small quantities can be made easily, resulting in more flexibility than in the lump-sum type of contract.

Many projects are constructed using both types of contract. The superstructure of a building, for example, usually consists of many known elements and readily adapts itself to a lump-sum contract. Details of excavation and foundation construction are often subject to change after construction starts and can be dealt with best on a unit-price basis. The construction of a dam or highway involves items of excavation, fill, and pavement placing that lend themselves best to the unit-price concept, but there may be structures such as small buildings or bridges that would be paid for as a lump sum. Particularly on large projects there may be elements of each type of contract within the general contract for the entire project.

3.3 NEGOTIATED CONTRACTS

The competitive contract, either the lump-sum or the unit-price type, is used on nearly all public works unless they be of an emergency nature. Private owners also use these types extensively since they are based on the same principles of free enterprise favored by most business in this country. Private owners may also negotiate directly with one or more favored contractors, and the resulting contract may take one of the following forms:

Cost plus a percentage of cost One of the oldest types of negotiated contract is one in which the contractor is reimbursed for all his costs and in addition is given a fixed percentage of that cost to cover his profit. Overhead costs may be covered by the percentage, or they may be reimbursed by the owner as one of the costs. A contract of this type permits construction to start before the plans are complete and may be an important factor when time of completion is important. Changes are easily made as the project progresses. Of disadvantage to the owner are the fact that the total cost of the project is completely unknown before the project starts, and the cost will often be higher than with some other type of contract.

In making a competitive bid, a contractor must anticipate some uncertainties and hazards and make allowance for them by including an amount for contingencies in his bid. By removing this risk from the contractor the owner hopes to have his project completed at a lower cost. If, however, there are more hazards present than the contractor might have anticipated, under the cost plus arrangement the owner must pay for them, resulting in a higher cost than with the fixed-price contract. Since the contractor's profit increases as costs increase, there is no incentive for economy on his part, and often an unscrupulous contractor will increase his apparent costs dishonestly. Since the contractor has no great interest in economy, his supervision of the job and the choice of equipment and men supplied to the job will not be of the caliber typical of a fixed-price job. Laborers, foremen, and job superintendents all have some degree of loyalty to the employer, and if they realize that the contractor has no need for their best efforts, efficiency will go down and quality and cost will both suffer.

For the contractor, the benefits of this type contract are obvious and can be enormous. He cannot possibly lose money, and the more inefficient he is, the greater will be his profit. Since he cannot lose, he may be willing to undertake any project, whether he has the competence and equipment or not. The success of this type of contract varies directly with the integrity of the contractor, and many satisfactory projects have been undertaken and completed in this manner. This form of contract is often used in emergency and repair jobs since there may not be enough time for negotiating any other type.

Cost plus fixed fee The outstanding defect in the cost-plus-percentage contract stems from the connection between cost and profit and the resulting tendency of the contractor to increase costs. In the cost-plus-fixed-fee contract, the owner, as before, pays all costs. In addition, he also pays the contractor a fixed sum of money rather than a percentage of the cost. The fixed fee may or may not include overhead costs, and its amount is determined by the estimated cost of the project, even though the cost may be relatively unknown before the project starts. The amount of the fee is negotiated between the owner and contractor or may even be the subject of bidding among selected potential contractors. The time involved in negotiating the fee may delay the start of the project somewhat, but construction can still start long before plans are complete. With the owner assuming all risks by paying all costs, the resulting construction may be at a lower cost than with a fixed price. Since profits are not related directly to costs, there is no incentive for the contractor to inflate costs deliberately. Prolonging the completion of the job to increase costs acts to decrease contractor profits, especially if he must pay overhead out of the fixed fee, so faster completion of the job usually occurs as compared with a cost-plus-percentage contract.

The unknown total cost of the project is a disadvantage to the owner, and the lack of incentive for economy may make for inefficiency and resulting poor workmanship. This type of contract has been favored by the U. S. government for many projects during wartime when delay in starting a job should be a minimum. Even with its well-known reputation for waste, the government has avoided cost-plus-percentage contracts with their many possibilities for poor but expensive work.

Cost plus fixed fee with bonus In many cases the time of completion of a project is of great importance to the owner. In the case of a manufactured product, for example, the first company that can market something new has a great advantage over its competitors, and completing the manufacturing facility as quickly as possible may be vital to the business. In such cases a contract for cost plus a fixed fee with a bonus for early completion might be used.

The owner and contractor agree on a target date on which the construction should be completed. For each day before that date on which the owner can use the completed facility, he pays the contractor a fixed, agreed-upon amount. For late completion the contractor may be required to pay the owner an amount based on the owner's loss at not having the use of the project.

Cost plus fixed fee with profit sharing Use of this scheme requires the owner and contractor to set a preliminary estimate, or target, on the cost of the work. After completion, any saving of cost is shared by owner and contractor in an agreed-upon proportion. The purpose of this type of contract is to give the contractor an incentive to minimize costs. In any event he will receive his fixed fee, but the possiblity of increasing his profit helps to increase efficiency, with resulting savings to the owner. Another advantage to the owner is that he has some idea of the total cost at the start of the project. With a preliminary target required, start of the project may be delayed somewhat, but work can still begin before plans are anywhere near completion.

Many variations of these four types of contract can be made in which features of profit sharing or bonus for early completion can be combined into one contract, or the idea of a fixed fee may be combined with the percentage of cost idea so that the fee is allowed to change somewhat as cost changes of a large magnitude occur.

Management contracts When a contractor is awarded a contract it is assumed that the work will be done with his own work forces even though a certain amount will be performed by subcontractors. Some contacts for highway work stipulate that at least half of the work must be done by the prime contractor. The situation in which a contractor, when accepting a contract, plans to have most of the work performed by subcontract without prior knowledge of the owner, is called brokerage and would not be allowed if the owner knew about it. Sometimes, however, an owner would prefer the work to be done by a contractor in whom he has a great deal of faith but who cannot accept additional work at this time. In this case the favored contractor might be given a management contract, giving him responsibility for supervising construction, letting subcontracts as needed, and ordering and paying for labor and materials. The manager would be chosen for his ability and integrity, even though the work would not be performed by his own personnel and equipment. The manager would not act as the owner's agent but would be an independent contractor, with the method of doing the work left to his own discretion.

Engineer–architect–management This type of contract was widely used during World War II in the rapid building of the military installations of all sorts that were necessary for the war effort. Using this arrangement, one firm of engineers or architects was given responsibility for the planning, design, and letting all the construction contracts for a given installation. The actual construction was all done by contract between the designer and the contractor. The architectural or engineering firm in this type of operation acts as an independent contractor, not an agent of the owner, and is responsible for all damages or harm to third parties. There are no contractual relationships between the owner and the contractor responsible for the actual physical construction. The form of contract between the owner and the architectural or engineering firm is usually of the cost-plus-fixed-fee type, with the owner taking little part in determining the contract for the physical construction.

Combined design and construction The "turn-key" or "package" job, widely used in Europe and South America for many years, results when one company is given the responsibility for designing and building a project. Often both the design and the construction forces are employed by the same company.

The "design and build" contracts that are beginning to be popular in this country are of this type, and often the contractual arrangements are between the owner and contractor only, with the design architect or engineer being either a member of the contractor's firm or chosen and paid by the contractor. In this country the architect or engineer responsible for design is usually an agent of the owner and inspects the construction to safeguard the interests of the owner. In his capacity as agent he is responsible for interpreting the plans and specifications, and although paid by the owner, his professional judgment is relied upon in disputes between the owner and contractor. This valuable contribution is lost in the package job, and the architect or engineer becomes either a captive or partner with the contractor, with nobody left to protect the interests of the owner. With a good contractor, this type of work can result in good and efficient work at a fair price. Many contractors and their trade organizations today are demanding professional status in the public eye and are urging acceptance of codes of ethics by members of the construction industry. Responsible action in this regard by the industry will certainly help in making this type of construction as popular and satisfactory here as it is abroad.

Joint venture Large construction projects often require the use of more equipment and other resources than a single company has available. A joint venture results when two or more companies combine their resources for a particular job. The result resembles a partnership between two or more companies but applies to that one job only. There should be an agreement between the companies defining the manner in which the personnel and management of the participating contracting companies will be used, how the project will be financed, and how the profits or losses will be shared. The agreement between the companies must often be approved by the owner and sometimes becomes a part of the prime contract between owner and contractors. Although a temporary partnership exists for one project, the members of the joint venture may be in fierce competition with each other in bidding for other jobs.

Dams and military installations were formerly typical of the types of projects large enough to warrant joint ventures, with the giants of the industry combining forces for a particular job. Today, however, we find small and medium-sized contractors joining forces to bid on jobs that require a variety of skills that formerly were considered by only the large firms. The joint venture thus makes it possible for a combination of small companies to compete in "the big time."

3.4 SELECTION OF TYPE OF CONTRACT

The selection of the type of contract to be used for a construction project is made by the owner, acting upon the advice of his engineer or architect and his legal advisor.

The use of competitive bids for awarding a lump-sum or unit-price contract is indicated when plans are complete and the owner is interested in securing the low price that competition in the open market will bring. He must be willing to furnish the inspection and supervision necessary to make sure that construction is in accordance with the plans and specifications. Local business conditions will control the amount or lack of competition available between contractors. The competitive bidding results in the type of contract that many businessmen are familiar with and use for many ordinary business transactions—not only construction contracts.

A negotiated contract should be used when construction should start before plans are complete or when the many unknown factors of the project make an accurate cost estimate impossible. When many changes in the work are expected and when inspection and supervision of the work cannot be done effectively, the negotiated type of contract should be used. The cost-plus-fixed-fee or cost-plus-percentage-of-cost contracts are the easiest to administer, especially when overhead costs are included in the fee, since the contractor merely submits his paid bills or vouchers for all his costs to the owner and is reimbursed for them. The fee can be paid at completion of the project or in installments agreed upon previously.

Contracts for work outside the United States usually contain incentive clauses holding out the possibility of large financial reward to the contractor. Work in a foreign land often involves transportation problems, labor shortages and the necessity for training native workers, housing and feeding problems, language and communication difficulties, speculation resulting from dealings with unstable governments and currencies, and many other conditions that might make it very difficult to fulfill a contract. A profit-sharing type of contract is often used in which the contractor submits a target estimate and receives a share of the savings over his estimate. His fee may be lowered somewhat if his cost exceeds his target.

3.5 RENEGOTIATION

Under usual contract law an agreement, once made, is valid and stands even though one party makes more profit or gets more of a bargain than originally planned. Contracts made with the federal government during time of war or other national emergency are often made in haste so that construction can be completed in the least possible time. Even though it is contrary to general contract law, all contracts amounting to a value of more than $100,000 dealing with war or defense materials or construction are subject to review to determine whether profits have been excessive or not, since public policy dictates that nobody should profit unduly from war or defense contracts.

This renegotiation of contracts started during World War II and later action by Congress has renewed it. During the war, many manufactured articles were paid for on a cost-plus-fixed-fee basis, with the amount of the fee determined by the cost of a prototype, a few manufactured articles, or by an estimate supported by insufficient facts. Some manufacturers reduced their costs greatly, to the point where the fee, as a percentage of cost, was really exorbitant. These cost reductions were often the result of astute management and use of engineering skill, and to penalize a manufacturer for his efficiency is unfair to him, but to allow him to make overly large profits is not in the public interest in time of national emergency, hence renegotiation to force the return of excess profits.

3.6 SUBCONTRACTS

Many specialized phases of work on a project are not performed by the prime contractor but are done on a subcontract basis. Employment of a subcontractor is usually by competitive bid, using either the lump-sum or the unit-price concept or a combination. The resulting contract is between the prime and the subcontractor.

In competitive contracts between owner and prime contractor the owner sometimes reserves the right to approve all subcontractors, but since the prime contractor is responsible for the work of his subs, this is not usually deemed necessary. In negotiated contracts, however, since the work done by the subcontractors is one of the items of cost, relations between prime and subcontractors are of importance to the owner, and these contracts should always have the approval of the owner.

CHAPTER 4

Contract Documents

A contract may be a very simple oral or written agreement, but the complexity of most construction projects demands a rather lengthy and complicated description of what is to be done and the mutual responsibilities of the two parties. Instead of trying to combine all necessary information into one document, it is the usual custom to incorporate all necessary information into several contract documents, with each becoming part of the total contract. The contract documents for a medium-sized building might consist of a hundred or more pages.

4.1 ADVERTISEMENT

The first contract document seen by the prospective bidder is the advertisement, or notice to contractors. The purpose of this document is to notify contractors that a contract is to be awarded and thus obtain the desired competition. The advertisement appears in newspapers, magazines, and trade journals having circulation among the the contractors in the construction area.

Contractors are often notified directly by mail that a contract is to be let in order to increase competition. The advertisement is designed to notify contractors that a bid is requested and gives enough information to allow them to determine whether they are interested or not.

4.2 INFORMATION INCLUDED IN ADVERTISEMENT

Advertising costs money, with the cost determined by the number of words or number of lines. The information given is therefore usually rather brief and contains such as the following: description of the work and approximate quantities of the items of work, name and address of the owner, time and place of bid opening, character of bids (whether unit price or lump sum), name of the architect or engineer, information regarding plans and specifications and any costs relating to procuring them, and information relating to bid and contract surety. The example to follow is fictitious and is simply an example of typical standard practice.

EXAMPLE OF ADVERTISEMENT

> Sealed bids for the construction of an apartment building in Amherst, Massachusetts, will be received at the office of John Brown Associates at 222 Main Street, Amherst, until 10:00 A.M. on Tuesday, September 19, 1974, at which time they will be opened and read publicly.
>
> The building, consisting of 20 one- and two-bedroom apartments, will be of two stories, brick veneer on standard wood-frame construction. All landscaping, walks, and drives will be constructed later by the owner.
>
> Contract documents may be examined at the office of the owner, Smith Realty Company, at 514 West Pleasant Street,

Amherst, and at the office of John Brown Associates, 222 Main Street, Amherst. Contract documents may be obtained at the office of John Brown Associates upon deposit of $10 for each complete set. The full amount of deposit will be returned within 10 days after the date of opening bids provided that documents are returned in good condition.

A bid bond of 10 per cent of the bid amount must be submitted with each bid and will be refunded for all but the three lowest bidders within 10 days after opening of bids. Performance and payment bonds equal to 100 per cent of the total contract price will be required.

The owner reserves the right to reject any or all bids and to waive informalities in any bid.

The bid for all excavation and substructure will be of the unit-price type, the bid for the remainder of the building to be of the lump-sum type.

4.3 INSTRUCTIONS TO BIDDERS

If the advertisement is successful in interesting a prospective bidder enough that he wishes to know more about the job, the next step for him is to examine the plans and specifications at one of the locations specified in the advertisement or to get a set of contract documents to study at his leisure. Included will be the Instructions to Bidders, or Information for Bidders. The purpose of this document is to make sure that all bidders have the same information on the unique features of the project. If the advertisement is not specifically made one of the contract documents, all information contained therein is repeated in the instructions, since there can be no assurance that all bidders have even seen the advertisement, and it is important that all bidders have identical information concerning the job.

4.4 INFORMATION INCLUDED IN INSTRUCTIONS

The information in this document is usually more detailed than that in the advertisement and pertains to features of interest in the actual bidding procedure as well as to unique features of the project itself. The purpose of this document is not only to help in compiling the bid but also to explain more fully than in the advertisement those features of the project that might cause him to become either more or less interested in compiling a bid.

Some of the topics often covered in the Instructions to Bidders are: a detailed estimate of quantities of the major parts of the work for unit price contracts, an exact definition of the scope of the work if a lump sum contract will be used, time of completion of the project, requirements concerning the bidders' qualifications, instructions to be followed in submitting the bid, disposition of late bids, procedure to be followed in awarding the contract, definition and disposition of bid and contract surety, withdrawal of bids, rejection of bids, responsibility of accuracy and interpretation of bidding information, and mention of any unique features of the project. For a small project only a few items need be mentioned, whereas many pages might be necessary for a medium- or large-sized building. The example to follow is meant to illustrate the wording used and some of the items included in a small- or medium-sized project.

EXAMPLE OF INSTRUCTIONS TO BIDDERS

1. *Interpretation of contract documents* Question concerning the meaning of or interpretation of any part of the contract documents must be submitted in writing to the architect at least 10 days before the receipt of bids. The architect will answer as addenda all questions received. These addenda will be furnished all bidders at least 5 days before the date for receipt of bids.

2. *Site conditions* Bidders should visit the site of the work to determine by inspection and inquiry pertinent local conditions such as location, labor conditions, and any other factors that might affect the cost of the work.

3. *Borings* Borings have been made at the locations shown on the drawings, and samples of material encountered may be examined at the office of the architect. The owner does not warrant or guarantee that the materials encountered during construction will be the same as those of the samples. Any interpretation of the results of the borings is the responsibility of the bidder.

4. *Preparation of bid* Bids must be submitted on the attached form. Bid prices must be filled in, in ink, using both words and figures. Unit prices will control in case of a discrepancy between unit price and amount. Changes and erasures must be initialed by the bidder. In any discrepancy between words and figures, the words will control.

5. *Signing bid* Each bid must contain the name, business address, and usual signature of the person making the bid. Partnership bids must list the names of all partners and must be signed with the partnership name by a partner or other authorized representative, followed by his own signature and designation or title. Corporate bids must be signed with the corporation name, followed by the signature and designation or title of the person signing. The address must be given for all partnerships and corporations.

6. *Receipt and opening of bids* Bids will be opened and read publicly at the time and place noted in the advertisement. Late bids will be accepted if such lateness can be proved to be caused by an abnormal delays in the mails or by a telegraph company. Such late bids, to be considered, must be received before awarding of the contract.

7. *Withdrawal of bid* A bid may be withdrawn in person or by written order at any time prior to the bid opening.

8. *Award of contract* The contract will be awarded to the

lowest responsible bidder as recommended by the architect. The contract will be awarded within 5 days after opening of the bids. Before awarding the contract the owner must be satisfied that the bidder has a permanent place of business, has adequate plant and equipment to do the work, has sufficient financial resources, and can submit a satisfactory performance record. The owner reserves the right to waive any informalities in any bid.

9. *Time of completion* The contractor shall start work within 10 calendar days after being awarded the contract and shall complete the project no later than 200 calendar days after the award. Liquidated damages of $50 per calendar day will be charged the contractor for each calendar day required after that date.

10. *Bid bond* Each bid must be accompanied by cash or certified check in the amount stated in the advertisement. All checks must be made payable to the owner.

4.5 PROPOSAL

The proposal is the contract document in which the bidder offers to furnish the materials and perform the work described by the plans and specifications and other contract documents, and for this he agrees to accept a given consideration. If the proposal is properly prepared it will contain the five essentials of a contract, so that if the owner accepts the proposal a valid contract has been made.

The proposal is often prepared by the architect or engineer, but its legal importance is such that an attorney is often consulted, or it may be prepared entirely by an attorney. The form of the proposal is often prescribed, or, better still, it is often furnished by the owner, to make sure that all bidders are proposing to do exactly the same things, thus making it easier to compare bids. Standard proposal forms are often used when the work is of a common type and there are few unusual features.

4.6 INFORMATION INCLUDED IN PROPOSAL

In order to make clear exactly what the bidder proposes to do, the following items are usually included in the bid form or proposal:

1. The price for which the bidder is willing to perform the work specified. If a unit price contract is to be used, the bid item will be listed along with the estimated quantities. The bidder will insert his unit prices and calculate the dollar amount for each item, and the total of these amounts will be his bid price. If the contract is the lump-sum type, only the total price will be given. Prices are usually given in both words and figures. In case of conflict, the words govern.

2. The time of starting the work and the time of completion should be stated, either as calendar dates or as a given number of days, either working days or calendar days.

3. Listing all contract documents, including any addenda that have been issued during the bidding period, completely defines the project, and such a list should be included in the proposal.

4. A statement that the site has been visited and that the contractor understands all plans and specifications is often included.

5. A list of subcontractors is sometimes required.

6. A statement that no fraud or collusion exists is often required. The collusion may be between bidders, or there may be illegal agreements between a bidder and a representative of the owner.

7. It should be stated that a bid bond in the form of cash, certified check, securities, or other form of surety accompanies the bid, and the bidder also states that contract surety will be furnished if he is offered the contract.

8. Sometimes a financial statement or record of performance on projects of a similar nature is required.

9. Name and signature of bidder are required. The authority for the signature is required if the bidder is a partnership or corporation.

EXAMPLE OF PROPOSAL

Proposal

To the Town of Leverett, Massachusetts, acting through its Board of Selectmen:

The undersigned hereby proposes to furnish all labor and materials necessary in connection with the construction of a bridge over Roaring Brook on Amherst Road for the sum of__________ (\$__________).

I hereby declare that I have examined the location of the proposed work, and also the plans, specifications, and the contract on file in the Office of the Selectmen; that I will furnish all materials, tools, and equipment, and do everything required to perform the projected work in accordance with the plans, specifications, and other contract documents.

I declare that this proposal is made without collusion with any other bidder and that no agent or officer of the Town is directly or indirectly interested in this bid.

Included with this proposal is ____________________
Bid bond, Certified check

in the amount of ____________________ (\$__________), which shall become the property of the Town of Leverett if I shall fail to execute a contract with the Town of Leverett within 5 days after being notified in writing that this proposal has been accepted by the Town.

I also agree to present to the Town, at the time of signing the contract, a performance bond in the amount of 80 per cent of the amount of this proposal.

If this proposal is accepted by the Town, I agree to commence work within 10 calendar days after signing the contract and complete the work within 30 days after signing the contract.

The undersigned hereby designates ______________________ as his business address.

Dated ______________________

____
Name of company or individual

By ______________________

____
Title of person signing

Bidder is requested to list work similar in character to the proposed work that he has performed and give business references that will enable the Town to judge his experience and financial standing.

4.7 AGREEMENT

The most important document from a legal standpoint is the agreement. It is sometimes called the contract, but since so many documents are included as contract documents, the agreement is a better term for this particular one. The form of the agreement can be standardized and used for many projects, or a unique document can be prepared for each project. The standard form prescribed by the American Institute of Architects has proved to be satisfactory and has been used on many building projects with good results. The form followed for nonbuilding projects is often more varied, usually being of a form developed by the firm of architects or engineers responsible for the design of the project, and followed on all the projects undertaken by that company. Many public agencies also have their own standard agreements which are used on all their projects.

4.8 INFORMATION INCLUDED IN THE AGREEMENT

The first part of the agreement is usually a short introductory paragraph which defines the parties, gives the date of the agreement, and states that each party agrees to what follows. The next part contains the elements of a contract and defines the work to be undertaken, the consideration to be paid, sometimes the method of payment, the time of completion, provision for liquidated damages, and a complete listing of the contract documents that describe and define the project. A final paragraph confirming the agreement and providing space for the signatures of the parties and witnesses, if necessary, completes the agreement.

There is no such thing as a standard procedure that applies to all construction projects and clearly spells out the duties and obligations of each party. Failure to specify what is expected of each party leads to many misunderstandings and often to litigation. The many contractual obligations may be included in the agreement itself, or they may be made a separate contract document (as in the American Institute of Architects form of agreement). If separate, they are usually referred to as "General Conditions" and may be followed by supplemental conditions relating to the particular job. General Conditions are usually standardized and can be used for all jobs of the same nature. Whether a separate document is used, or the information is included in the agreement, the same sort of information is given. Many of the clauses or articles might seem unnecessary or obvious, but they are included to ensure that each party knows exactly what his obligations are. The more complex a project is, the more complex are all the contract documents, and certainly a small construction job does not warrant the detailed legal papers that are so necessary for large jobs. The General Conditions used in the standard form recommended by the American Institute of Architects consist of 14 articles designed to cover the problems encountered in most building construction.

The following example incorporates the necessary general conditions in the body of the agreement, thus requires no separate document for them. The example is rather brief and its purpose is to show a typical document, not to explore all the possible areas of misunderstanding between owner and contractor.

EXAMPLE OF AGREEMENT

Agreement

This agreement, made this __________ day of __________, 19__________, by __________, hereinafter called the Owner, and __________, hereinafter called the Contractor, witnesseth that the parties for the consideration hereinafter named mutually agree as follows:

Article I. Scope of Work The Contractor shall provide and furnish all the labor, materials, plant, and equipment necessary to complete all work required for the construction of __________, all in accordance with the plans and specifications, including all agenda, prepared by __________, which plans and specifications are hereby made a part of this contract; and in accordance with all other contract documents mentioned herein, which are a part of this contract.

Article II. Time of Completion The work to be performed shall be commenced on __________ and shall be completed no later than __________. If the Contractor fails to complete the work before the completion date given above, the Owner shall deduct from payments due the Contractor the sum of __________ per day for each calendar day of delay. This sum is agreed upon not to be a penalty, but as fixed and liquidated damages for each day of delay.

Article III. Contract Sum The Owner shall pay to the Contractor for the performance of this contract the sum of __________, the said amount being the amount stated in the Proposal.

Article IV. Contract Documents This contract consists of the following separate component parts, each of which is fully a part of this contract as is hereto attached:

1. Plans and other drawings numbered ________
2. Specifications
3. Advertisement
4. Instructions to bidders
5. Addenda numbered ________
6. Proposal
7. This document

Article V. Performance Bond Within 10 days after signing the agreement the Contractor shall furnish a bond, using the attached form, in the amount of 80 per cent of the contract sum as security for the performance of this contract.

Article VI. Contractor's Insurance Before commencing work the Contractor shall obtain, and maintain during the life of this contract, sufficient insurance to protect him from claims under workmen's liability and other employee benefit acts, from claims for bodily injury and death, and from claims for damages to property that may be caused directly or indirectly by operations within the terms of this contract. The Contractor shall furnish the Owner with certificates of insurance. Minimum amounts of coverage shall be established by the Architect.

Article VII. Owner's Insurance The Owner will maintain for the life of this contract insurance to protect himself from any contingent liability for damages that may be caused by operations within the terms of this contract. The Owner will maintain fire and extended coverage insurance in the amount of 100 per cent of all work in place and all materials stored on the site.

Article VIII. Laws, Regulations, and Permits The Contractor shall comply with all laws, ordinances, and regulations affecting the work in any manner and shall protect the Owner from any claim arising from violation of any such law, ordinance, and regulation. The Contractor shall obtain, and the Owner will pay for, all permits and licenses required for performance of the contract.

Article IX. The Owner will establish property lines, base lines, and bench marks as shown on the drawings. The Contractor shall establish all other lines and grades required for performance of the contract.

Article X. Payments On or before the 10th day of each calendar month the Contractor shall submit to the Architect an estimate of the value of work performed and materials stored on the site during the previous month. Upon receipt of the Architect's approval of this estimate and no later than the 20th day of the month, the Owner will make a progress payment in the amount of 90 per cent of the estimate. Upon substantial completion of the project the Owner will make a payment sufficient to make the total payments equal to 90 per cent of the total contract price. Upon completion of all the work included under the terms of the contract, the Architect will issue a certificate of completion and the Owner will pay the Contractor the unpaid balance of the contract sum within 30 days.

Article XI. Duties and Authority of the Architect The Architect shall be the person or his agent, appointed by the Owner, and authorized to perform the duties assigned to the Architect. The Architect shall have access to all work in progress and shall have the right to reject any materials and workmanship which, in his opinion, do not conform to the plans and specifications. All rejected work shall be corrected without charge to the Owner. Decisions of the Architect concerning the meaning of any of the contract documents and his decisions given in cases of errors or discrepancies or omissions in any of the contract documents shall be binding upon both parties.

Article XII. Extension of Time Delays caused by extra work, acts of any government, extreme weather conditions, or other acts beyond the control of the Contractor may entitle the Contractor to extra time to complete the project if the Architect so certifies in writing.

Article XIII. Changes in the Work The Architect may, by written order to the Contractor, make changes in material or dimension of the work. If such changes make the work more expensive, the Owner will pay the Contractor the cost of labor, the cost of materials, a fair rental price for all equipment, and 15 per cent of the sum of these for overhead, profit, and superintendence caused by such changes. If changes make the work less expensive, the Owner will deduct an amount calculated in the same manner from the contract sum. All additions or deductions arising from changes in the work must be approved in writing by the Architect.

Article XIV. Owner's Right to Terminate Contract The Owner reserves the right to complete the work as agent for the Contractor if the Contractor should abandon the work for a period of 15 days or more, or if he should become bankrupt, or if the contract should be assigned or sublet without permission of the Owner, or if the Contractor should violate substantial provisions of this contract. The Owner shall give the Contractor 7 days' written notice before taking possession of the materials, tools, and equipment and proceeding to complete the work himself, either by direct use of these materials, tools, and equipment or by use of other contractors. If the expense of completing the work exceeds the unpaid balance of the contract sum, the Contractor shall pay this difference to the Owner. If the reverse is true, the excess shall be paid to the Contractor.

Article XV. Contractor's Right to Terminate Contract The Contractor reserves the right to terminate the contract and recover payment for all work completed or in progress and receive payment for any loss sustained should the Owner fail to make payment within 10 days after it is due, or if the work should be stopped by an act or neglect by the Owner for at least 10 days. Before terminating the contract, the Contractor shall give the Owner 10 days' written notice.

In witness whereof, the parties have executed this contract as of the day and year first written above.

________________ Company

By______________

________________ Contractor

By______________

Witness:

____

CHAPTER 5

Drawings

Drawings and specifications, although classified as contract documents, are prepared by the designer of a structure or other project, are of a technical nature, and therefore will be considered separately from the other contract documents. The previously described documents are often prepared by legal personnel; although they contain some descriptive information concerning the project, their main function is to set up the essentials of the contract, protecting the owner, and being fair to bidders or the contractor at the same time. The function of the drawings and specifications, however, is to explain what is to be done and how it is to be done in language that can be understood by the contractor.

5.1 PURPOSE OF DRAWINGS

Engineering drawings is the graphic language used universally in industry to give the information necessary to build—whether it be a structure, a small machine part, or the machine itself. The most efficient way to describe a shape, to locate details, to indicate dimensions, and to show the relationships between parts is usually by means of a drawing. The oft-quoted saying, "One picture is worth a thousand words," is very often a gross understatement. The drawings used in construction must show the contractor what he is expected to do—the location, size, shape, and any special features of the work. To illustrate the superiority of a drawing over a verbal description, one can visualize the complexity of trying to explain in words how a simple object such as a wooden chair can be built. The verbal description of the size and shape of each part and the relationships between the various parts and how they are fastened together could easily fill several pages, while the same information could be given in two or three simple drawings.

5.2 ORTHOGRAPHIC PROJECTION

The purpose of a drawing is to convey sufficient information so that the object can be built. All dimensions and the locations of all details must be shown. We live in a three-dimensional world, yet the draftsman must show all three dimensions on a two-dimensional sheet of paper. Pictorial drawings and photographs can be used to a limited extent in showing the size and shape of an object, but some of the dimensions are distorted, and much detail, especially interior detail, cannot be shown clearly.

To overcome these shortcomings, orthographic projection is often used. If projection lines from an object are imagined to intersect a plane at right angles to the plane, the resulting image on the plane is an orthographic projection of the object. The same result can be obtained if we imagine that the lines of sight from an observer's eye to the object are parallel to each other, thus lines of sight from the observer and projection lines from the object to the image plane mean about the same thing.

Orthogonal projection makes use of three mutually perpendicular planes. These planes most often show the top, front, and right or left side of an object, and very often these three views—or sometimes only two of them—are sufficient to describe the object.

Top, front, and side views are basic views for small objects such as machine parts, but for buildings, dams, highways, and other large objects a different and more precise terminology is used, based on the direction of the lines of sight or projection lines:

1. A plan view—or simply plan—is a view that uses vertical lines of sight or projection lines. The top or bottom view of an object is thus a plan. A map shows a top view of the surface of the earth and is also a plan.

2. Elevations or elevation views are drawn using horizontal lines of sight. Elevations may be drawn from the front, back, or any side.

3. Elevation views that are neither front, back, or side elevations are sometimes called auxiliary elevations. They utilize horizontal lines of sight, but the image plane is not one of the principal ones.

4. Sectional views show what is left after the object has been cut by an imaginary cutting plane and part of the object removed. A sectional view may be drawn from any type of view—plan, elevation, another sectional view, etc.

5. Auxiliary views using lines of sight that are neither vertical nor horizontal are often used to show the true size and shape of an inclined surface.

5.3 ARCHITECTURAL DRAWINGS

A typical set of drawings for a building of medium size would consist of perhaps 20 or 30 drawings, often on 24- x 36-inch sheets, and would contain several types of drawings:

Site plan This drawing is meant to locate the structure on the building lot, and gives the locations of important points of the building with respect to property lines, highways, or other easily located points. Bench marks and other reference points, existing drainage and other utilities, locations of boring and test pits and any information available concerning subsurface conditions, as well as the outline of the proposed structure and any existing structures are usually in the site plan.

Topographic map A separate plan showing the existing and final elevations by means of contour lines may be necessary if it is impossible to include this information on the site plan. Boring logs showing the soil conditions encountered at various elevations may also be included here if they are impossible to show on the site plan.

Plan views A plan view of each floor, including the roof, is usually given. The foundation plan shows features in the plane of the foundation, such as elevator pits, drainage trenches, and the like, as well as the details of the footings, piles, foundation walls, and anything else required to support the building. The plan views for the various floors of the building show locations and sizes of walls and partitions and sizes and locations of all door, window, or other openings.

Elevations An elevation view of each side of the building is required to show the external appearance of the structure, floor and finish grades, ceiling heights, and other information pertaining to the use of building elements in the external walls.

Sections Many sectional views, usually to a larger scale than the previous views, are required to show the construction of exterior and interior walls, the construction details around doors and other openings, and much other information impossible or difficult to show on the other drawings. These sections may be longitudinal or cross sections of the entire building, or they may be of smaller size to show more clearly some detail of construction.

Detail drawings In many buildings there are some unusual features, architectural or mechanical, that require further explanation. Detail drawings, usually to a rather large scale, are necessary to describe these unusual or complicated features.

Structural Separate drawings showing the structural elements of the building are sometimes necessary, particularly for steel frames. On these drawings are shown the size and position of all columns, beams, girders, and trusses, often including details of connections when steel or timber construction is used. Structural steel and reinforcing steel for concrete structures are cut to size, holes are punched, and bends are made as required in fabrication plants. Identification numbers are painted on this steel before it is shipped to the construction site, and the structural drawings must show the location of these numbered elements.

Mechanical The drawings used by plumbers, electrical workers, and heating and ventilating workers are often lumped together and called "mechanical" drawings, even though separate drawings are prepared for each of these classes of workers. On them are shown size and location of all piping, valves, pumps, wiring, electrical switches and fixtures, and any other information needed by the several trades. Simple plan views show much of this information and much use is made of conventional symbols. Complicated piping is sometimes best shown by isometric or other pictorial drawing.

5.4 HIGHWAY DRAWINGS

Highway construction consists of earthwork, grading, paving, and placement of drainage appurtenances; consequently few types of drawings are required, although the total number of drawings may be very large. A typical set consists of the following:

General plan The purpose of this drawing is to show the entire highway, often including a profile, on one sheet. Little detail of the proposed construction can be shown because of the small scale of the map, but some topography, existing structures and roads, railroads, streams, and important property lines are included. Enough is shown to give an idea of the general location of the construction and some idea of the scope of the work.

Construction plans These plans or top views, often to a scale of 1 inch to 20 feet, can include enough detail to be useful in construction. The centerline of the highway and its stationing, limits of paving, sidewalks, and shoulders are shown. Limits of the right of way and property lines and names of abutters, all existing and proposed structures, and size and location of drainage pipes and appurtenances are shown.

Profiles On the same sheet with the plans or on separate sheets are drawn profiles of the centerline of the highway. These drawings are sectional views that show the height of every point on the centerline above a datum plane, usually mean sea level. In addition to showing elevations of the centerline, these drawings also show elevations of streams and other bodies of water, elevations of the existing ground, and all drainage features of the road. The vertical scale is often exaggerated or magnified with respect to the horizontal by a factor of 5 or 10 or more in order to better show the many surface irregularities.

Cross sections A profile is drawn using an imaginary cutting plane parallel to the centerline. By using cutting planes perpendicular to the centerline the cross-sectional elevational views are drawn. These are usually drawn for each 50-foot station along the centerline, but additional ones are used for points of curvature of the road, points where two or more roads intersect, and any other place where more detail is required. These views show elevations of existing land to the limits of the right of way, and have the proposed construction superimposed. Shape, thickness, and width of paving and shoulders are shown, as well as shape and thickness of subgrade, sidewalks, and curbing. Areas of cut and fill can be determined for each station, and from these the total earthwork requirements for the project can be determined.

Details Unusual features may require the use of detail drawings, which can be plans, elevations, or sectional views drawn to a large scale. Manholes, catch basins, details of reinforcing steel, fencing, and signs often require separate drawings. Since many of these are the same throughout a state or other geographical division, they can usually be described by references to a standard drawing, thus do not have to be drawn for each highway project.

5.5 DRAWINGS FOR HEAVY CONSTRUCTION

The drawings to be expected in various heavy-construction practices are a mixture of the types used in buildings or for highways. The drawings for an airport, for example, would be very similar to those used for a highway, since airport construction would consist of similar operations to those used in road building. Airport structures would require the same types of drawings as those used in a building project. The drawings for dam construction, also, would be much the same as for roads, since the operations would consist of much earth moving and building a structure of slowly varying cross section on an excavated area. The quantity of work is great, but the construction makes use of a relatively few crafts compared to the number involved in building construction.

CHAPTER 6

Specifications

6.1 PURPOSE OF SPECIFICATIONS

It is commonly stated that drawings tell what is to be done and specifications tell how it is to be done. Specifications supplement the drawings and help make clear to the contractor what is expected of him. Much information can be given on drawings, but many of the details of workmanship and material to be used cannot be described graphically.

A small concrete sidewalk can be shown very easily on a drawing, but the length, width, thickness, location, and the fact that it is to be placed on a layer of compacted gravel are about all that can be shown on the drawing. Many questions concerning the walk must be answered: What is meant by gravel? Should a sieve analysis of the gravel be made? What degree of compaction of the gravel is required? Concrete of what strength should be used? Are we concerned with the ingredients of the concrete? Should the concrete be mixed or deposited or cured in any special way? Do we care about the material or the placing of the forms? Are construction or expansion joints either necessary or desirable?

What finish is expected for the wearing surface? What variations from the given dimensions will be tolerated? Will the contractor repair any defects that appear in a given time period? Answers to these and other pertinent questions must be answered in the specifications if the contractor is expected to do the job to the owner's satisfaction. It should be obvious that if this number of questions arises on something relatively simple such as a sidewalk, in a complicated structure the specifications must be rather voluminous and must be designed to anticipate many areas of potential disagreement between the contractor and owner.

6.2 GENERAL SPECIFICATIONS

The first specifications encountered, the general specifications, are usually nontechnical in nature and apply to the project as a whole. In form they are very similar to the general conditions of the contract and sometimes repeat information already given in that document. There is no clear line of demarcation between the contract and the specifications as far as content is concerned, and the inclusion of a clause in one or the other of these documents is a matter of personal preference.

The general conditions of the contract are prepared by legal personnel primarily interested in protecting the interests of the owner. The general provisions of the specifications are prepared by technical personnel to describe the job and protect the owner's interests also; thus the wording and subjects contained in the two documents are similar. No attempt will be made here to include all the various clauses that might be included as general specifications, but those listed in the next paragraph seem to be typical of many jobs.

6.3 TYPICAL GENERAL SPECIFICATION CLAUSES

Location and description of work to be done This may be covered in some of the other documents, but a more detailed description may be included here.

Sanitary facilities provided for workmen These are usually supplied by the contractor, but a detailed description of their number and location and the responsibility for their maintenance should be given.

Materials supplied by the owner A listing of the materials and equipment, when they should be delivered, where they should be stored, and by whom they should be installed and tested is often necessary.

Utilities Provisions for temporary heat, water, and electric power must be made, as well as a clear understanding of who is responsible for paying for them. A listing of steam pressures, telephone lines available, and locations of water and electric power near the site will help the contractor in bidding and planning the job.

Storage Space for storage of materials and equipment is often of great importance, especially when the job site is in a crowded or metropolitan district. Obstructing traffic on highways or waterways, or interfering with the property of others, will not be tolerated.

Cooperation with others On large projects there may be more than one prime contractor, there may be many subcontractors, and there may be work being done by the owner—all concurrently—and the entire project must be planned in such a way that no conflict arises between the various workers. This is particularly important when liquidated damages are charged for late completion. Cooperation between the various contractors using the same excavations for wiring and piping, for example, should be defined and enforced.

Definitions and abbreviations Such expressions as "as directed," "or approved equal," and "as permitted" obviously should be defined either in the specifications or the contract. Not so obvious is the need for defining such terms as "day," "rock," "boulder," "ton," "earth," and "sea level." Such terms do not mean the same thing to everyone, and a clear definition of words subject to more than one interpretation can save litigation and other future trouble. Reference is often made to specifications and other publications of trade organizations, and such abbreviations as A.S.T.M. for American Society for Testing Materials and A.S.A. for American Standards Association should be listed.

Wage rates The wages paid to various classes of workmen are often specified, especially in public works. The actual wages to be paid are listed rather than a statement to the effect that workmen are to be "paid the commonly accepted wage rate for the locality."

Sequence of work In some cases it is important for certain phases of the project to be completed before others, and this should be noted. In general, however, the scheduling of the project is the province of the contractor, often subject to the approval of the engineer. The sequence of work should not be specified in enough detail that the contractor can be considered an agent of the owner. All specifications should be such that the independent contractor status is preserved.

Drawings Responsibility for preparing and provision for approving detail and shop drawings for temporary staging, concrete forming, ornamental metal work, and the like should be specified. Fabrication, approval, and ultimate ownership of samples and models must be provided for.

Protection of the work The provisions that must be taken by the owner or contractor to protect the work against vandalism, adverse weather conditions, or any other enemies must be clearly spelled out.

Offices The size, location, and ownership of any offices required for the contractor and the owner's representative must be noted.

The minimum furniture and other equipment, type of lighting, and method of heating are often specified.

6.4 SPECIFICATION WRITING

The preparation of specifications is a very exacting task, partly an art and partly a science. The specification writer must have a thorough knowledge of the entire job, must understand the properties and behavior of the materials to be used, and, especially, must be able to express his ideas clearly. The specifications are used by engineers, architects, material suppliers, supervisors, and many classes of workmen and must be understood and interpreted the same by all. Since the complete description of the project is contained in the drawings and specifications, the importance of good specifications cannot be overemphasized.

Specifications are not a showcase for the author's erudition nor should they contain legal terminology. Clear and concise writing is necessary. Complex sentence structure should be avoided, and the most important punctuation mark is the period. Short sentences, each dealing with one thought or idea, are used. The extensive use of pronouns often leads to more than one possible meaning for a sentence and should be avoided. Complete sentences, correct grammar, and the use of common words and phrases are required.

Great care should be exercised in the selection of words. Words with more than one meaning and words with a very general meaning should be avoided. Slang expressions, colloquialisms, and trade and technical expressions should not be used. Repeating one word several times in one sentence may be unacceptable for most writing, but it is often done in specifications. In conversation and general writing extensive use is made of synonyms, but since there are very few words in the English language whose exact meaning can be duplicated by another word, the use of synonyms should be avoided in preparing specifications. "Amount" and "quantity," "any" and "all," "must" and "shall," and "either" and "both" are a few examples of words that are not equal in meaning and should not be used interchangeably.

Specification sentences should be as brief as possible without losing the meaning of the sentence. Unnecessary words should be omitted, and if a word can be used to express the same thought as a phrase, the word should be used. Only those features that have a bearing on the end result should be specified. Reasons for specified requirements are never given. Information contained on the drawings or in other contract documents should not be repeated in the specifications. Both the importance and the complexity of an operation should have a bearing on the length of the specification. The materials and workmanship required for the concrete in a reinforced concrete building, for example, are important enough to be specified in great detail and many pages should be devoted to them, whereas the concrete to be used in a small sidewalk can be described rather briefly.

Specifications should be as definite as possible. Such expressions as "best quality," "as directed by the Architect," and "or equal" should be avoided if possible, since they do not tell the contractor what is expected of him. Some decisions must always be made in the field, but adequate planning will eliminate many of these. Specifications should, as much as possible, inform the contractor in advance what the job requirements will be. Indefinite and indeterminate requirements can often be interpreted so as to be unfair or to impose a hardship on the contractor, and this is anticipated by the contractor in making his bid. Vague requirements thus act against the best interests of the owner and lead to poor relations and possibly litigation between contractor and owner.

Standardized specifications for the various components of a project are often used with satisfactory results. Trade and other organizations, as well as government agencies, have standard specifications for items of equipment, machinery, and construction components. Rather than making new specifications for these, the standard specifications can be used by reference to them.

6.5 TECHNICAL SPECIFICATIONS

As indicated in the section on the purpose of specifications, there is much necessary information concerning a construction project that cannot be shown on the drawings. This detailed group of instructions must be given in the technical specifications. Each type of construction or each trade should have its own section for guidance and for describing exactly the result expected or the materials and methods to be used.

There are three types of technical specifications that can be used; the type to be used depends upon the particular operation. Many types of construction can best be described by specifying the quality of materials and workmanship expected in their use. This type, that is, specifying materials and workmanship, is very commonly used for building components, excavation, earthwork, piping, and similar projects in which both the materials and workmanship can be inspected and controlled. The adequacy of the finished product is then the responsibility of the designer. The contractor is responsible only for making the parts and assembling them as directed.

Technical specifications are usually arranged in sections, one section devoted to a particular trade or material. A subcontractor or materials supplier should not be required to read all the specifications, so all he needs to know should be in the particular specification relating to his trade or material. Regardless of the type of specification, the first paragraph usually refers the reader to the general specifications and any other contract documents that might affect him, notifying him that those documents are a part of the specification. The second paragraph contains the scope of the work and often states exactly where the work to be described is located, either physically or on the drawings. The remainder of the specification is devoted to a description of the work.

6.6 SPECIFICATIONS FOR MATERIALS AND WORKMANSHIP

For most construction materials there are several factors that can be used to determine quality. The use that will be made of the material will often dictate which of these properties should be used in the specification. For some materials the color, texture, or appearance are most important and they should be specified rather than strength or chemical or physical composition. Thermal and electrical properties, durability, corrosion resistance, hardness, waterproofness, and physical or chemical purity are characteristics of some other materials that should be used in their specifications. Any of these factors used in specifying quality should be factors that can be controlled and tested. Any special precautions used in storage, transportation, and installation of materials should be specified. Enough information to ensure the required quality should be given, but no more. Giving too much information sometimes leads to an impossibility, and can also change the status of the contractor to that of an agent of the owner. Specifying both the chemical composition and the electrical conductivity of wire, for example, might lead to impossibility of performance on the part of the contractor, thus making the owner completely responsible for the performance of the finished product. Paragraphs on workmanship contain details on connections, how well-finished parts must fit, smoothness of surfaces, tolerances, and other details pertaining to the skill of the workmen. Also specified are methods and order of assembly if needed.

The concrete in a reinforced concrete building offers an example of a material important enough that it should be described in some detail. The quality of each of the ingredients is important and must be specified. The method of mixing and transporting must be carefully described since they have a bearing on the finished product. Both temperature and moisture while the concrete is curing must also be specified. The minimum time before stripping forms and any surface treatment after placing the concrete must also be noted. The testing program used to determine the quality of the concrete should be noted as well as any necessary interpretation of test results.

The example to follow is typical of the concrete specifications necessary for a medium-sized building. The length and complexity of the specifications are proportional to the quantity and complexity of the work. In a relatively small job, much can be left to the discretion of the architect or engineer on the job, and even though some of his demands may be unreasonable and arbitrary, they will have little effect on the overall cost of the project. In a large project, however, the specifications must be thorough enough that the contractor will know before starting exactly what is expected of him so that his procedure and costs can be anticipated and plans made accordingly.

Unfortunately, the quality of concrete cannot be judged until tests have been made on samples made at the time of placing, the most important being the 28-day test. By that time, if the test results indicate inadequate quality, enough of the project may have been completed that requiring the contractor to replace any weak material could be an extremely costly solution to the problem. In the example the contractor is given alternatives, and if all else fails and the material is still under the required strength by a small margin, his payment is reduced. If the test results show very inferior material, however, the faculty concrete should be replaced. Recent trends in specifications for concrete seem to follow the reasoning that the contractor's payments should be reduced for somewhat understrength material, rather than a flat statement that the material should have a certain strength and nothing less than that will be tolerated.

For small jobs transit-mix trucks are used almost exclusively, and in the example only this type of mixer is mentioned. For very small jobs hand mixing can be used, and for large jobs a stationary plant can be set up, but these possibilities are not included.

Concrete used for various parts of a structure do not usually require the same design strength. Often that used for footings, foundation walls, and mass concrete such as is used for dams will be designed for relatively low strength, such as 2,000 or 2,500 psi. The concrete used for columns, beams, girders, and floor slabs often must have strengths of 3,000 or 4,000 psi. Thin shells and domes often use concrete testing at 5,000 psi or more.

The maximum size of coarse aggregate is also governed by the requirements of the job. For mass concrete certainly, relatively large aggregate should be allowed. The aggregate used in concrete for thin shells, small beams, thin floor slabs, and structural elements containing large quantities of reinforcing steel should be much smaller in order that the concrete may be worked around the steel adequately.

For purpose of simplicity the example uses concrete of only one strength and with one size of aggregate throughout.

EXAMPLE OF MATERIALS AND WORKMANSHIP SPECIFICATION

Sec. 4. Reinforced Concrete

4-1 General Requirements

a. The printed form of the contract and Section 1 of these specifications are hereby made a part of this section of the Specifications.
b. Standard tests of the cement, aggregates, water, and concrete will be made by the Owner. The Contractor shall furnish the material for all samples and shall provide all labor necessary for obtaining samples. The Contractor agrees to accept as final the results of tests secured by a testing laboratory engaged by the Owner.

4-2 Scope of the Work

The scope of the work in this section includes footings, foundation walls, beams, girders, columns, slabs, and all other concrete work shown on Drawings S-1 through S-5.

4-3 Cement

Cement shall be American-made portland cement, free of water-soluble materials that will cause efflorescence on exposed surfaces. Cement shall be the same brand throughout the work. Cement shall meet the latest ASTM Standard Specification for Portland Cement, Designation C-150.

High-early-strength cement may be used with the written permission of the Engineer.

Admixtures will be permitted only with the written permission of the Engineer.

4-4 *Water*

The water for concrete shall be clean and free of injurious amounts of acid, alkali, organic matter, or other substances that might affect the strength of the concrete.

4-5 *Fine Aggregate*

Fine aggregate shall consist of inert natural sand conforming to the latest ASTM Standard Specification for Concrete Aggregates, Designation C33. It shall not contain more than a total of 5 per cent by weight of the following: shale, silt, and structurally weak particles. Grading of fine aggregate shall conform to the following sieve analysis:

Sieve Size	*Per Cent Passing (by weight)*
No. 4	95–100
No. 16	50–80
No. 50	15–30
No. 100	3–8

4-6 *Coarse Aggregate*

Coarse aggregate shall consist of clean, well-graded gravel or crushed stone conforming to the latest ASTM Standard Specification for Concrete Aggregate, Designation C33. It shall not contain more than a total of 5 per cent by weight of the following: shale, silt, and structurally weak particles. Grading of coarse aggregate shall conform to the following sieve analysis:

Sieve Size	*Per Cent Passing (by weight)*
2 inches	100
1½ inches	94–100
¾ inch	50–70
⅜ inch	10–35
No. 4	0–5

4-7 *Concrete Quality*

The proportions of the ingredients of the concrete shall be such that the minimum allowable 28-day compressive strength shall be at least 3,000 psi, and the 7-day strength shall be 2,000 psi. The maximum allowable net water content shall be 6 ½ gallons per sack of cement. The minimum amount of cement per cubic yard of concrete shall be 5 ½ bags. The maximum allowable slump as determined by the ASTM Standard Method of Slump Test, Designation C143, shall be 4 inches. The proportions of concrete ingredients shall be submitted to the Engineer for approval at least 30 days before placing, and no concrete shall be placed without written approval of the proportions by the Engineer.

If concrete of the required quality is not being produced as the job progresses, changes in materials or proportions may be ordered by the Engineer. The Contractor shall make the required changes at his own expense.

4-8 *Mixing Concrete*

Mixing of the concrete ingredients shall be done in accordance with ASTM Standard Specifications for Ready-Mixed Concrete, Designation C-94.

Mixing water shall be added and mixing started immediately after the truck is charged with the dry ingredients.

Mixing shall be continued for at least 10 minutes and for at least 5 minutes after all the water has been added.

Agitation of the mixture (at less than mixing speed) shall continue until discharge of the mixture.

No batch mixed longer than 1½ hours shall be used.

Retempering of concrete that has partially set will not be permitted.

4-9 Placing Concrete

Before placing concrete the Contractor shall remove all debris, snow, ice, and water from the space to be occupied by concrete, and shall oil or wet forms thoroughly with water. Placing of concrete shall be continuous until reaching the end of the section or a construction joint. The concrete shall be placed so as to maintain a top surface approximately level. No concrete shall be placed on concrete partially set such that planes of weakness can form.

Concrete shall be worked into the forms and around the reinforcement using mechanical vibrators or hand spading. The working should produce a dense, homogeneous mixture without permitting segregation of the dry ingredients. Temperature of the concrete shall be between 65 and 100°F.

4-10 Curing of Concrete

Concrete shall be kept continuously moist for a minimum of 5 days.

Concrete shall be kept at a minimum temperature of 70°F for not less than 3 days or at a minimum of 50°F for 5 days.

If high-early-strength cement is used, curing time and temperatures may be modified by the Engineer.

4-11 Stripping of Forms

Forms may be removed from columns, beams, and slabs after 600 day-degrees, and from walls and other vertical surfaces after 150 day-degrees. A day-degree is defined as 1 day multiplied by the average air temperature in degrees Fahrenheit. If high-early-strength cement is used, forms may be removed as directed by the Engineer.

4-12 Testing of Concrete

Tests for the strength of concrete shall be done in accordance with the latest ASTM Standard Method of Test for Compressive Strength of Molded Concrete Cylinders, Designation C39. Two specimens shall be made for each test at a given age, and at least one test shall be made for each day's pour. The specimen containers will be furnished by the Owner and all sampling and testing will be done under the direction of the Engineer. The Engineer may require additional tests during the progress of the work.

If the average strength of any set of 7-day tests is less than 2,000 psi, the Engineer may require the Contractor to conduct further curing of the portion of the structure represented by the test specimen.

If the average strength of any set of 28-day test specimens is less than 3,000 psi, the Owner may take core samples from the portion of the structures represented by the test specimens. If the average strength of the core samples is less than 3,000 psi, the Engineer may require the Contractor to conduct further curing for 10 days, after which additional core samples may be taken by the Owner.

> If the average strength of the additional core samples is less than 3,000 psi but at least 2,500 psi, the Owner will deduct twenty-five cents ($0.25) from payments due for each 1 per cent deficiency in strength multiplied by the number of cubic yards of concrete in the understrength portion of the structure.
>
> If the average strength of the additional core samples is less than 2,500 psi, the Engineer may require the Contractor to strengthen or replace the portion of the structure represented by the samples.

6.7 SPECIFICATIONS FOR PERFORMANCE

In the performance type of specification the supplier or contractor is instructed to supply an article or piece of machinery capable of performing a specific task under given conditions. The choice of the article's manufacturer or its design is then left up to the supplier, and all responsibility for its installation, testing, and guarantee belongs to the supplier.

The purpose of performance specifications should be to describe the article sufficiently that it will be adequate for the purpose intended, but not in such detail that the product of only one manufacturer will meet that description. Many details of design not important in operation are sometimes included, resulting in an exclusive or restrictive specification. Such specifications are considered unethical, since they restrict the contractor unduly in his choice of product, and by restricting competition, act against the best interests of the owner.

The information given in the specification is often arranged to correspond to the sequence of events followed by the supplier in purchasing, installing, and testing the article specified, as follows:

Name and description Limiting dimensions and weights, recommended or necessary materials, and operating conditions such as temperature, pressure, and humidity should be noted. Fuel or other power requirements and switches and other controls should be described if necessary, but only enough information should be given to ensure correct operation of the article.

Existing conditions If machinery is being specified, the existence of fuel or other power is important, and if the availability of gas, steam, or air in addition to fuel is important, this, too, should be noted. Extreme or unusual temperature, pressure, and humidity conditions, if present, are important to the supplier and should be described.

Installation Responsibility for the installation may rest with the supplier, the owner, or a third party and must be clearly designated. Any unusual features of the installation, removal of old or other equipment, and the responsibility for any structural changes must be made clear. Connections of machinery to existing services may be made by the supplier or the work force employed by the owner. In either case the responsibility should be made clear.

Testing Determining if the article is suitable for its intended purpose is usually accomplished by some sort of test. In some cases simply noting that the article fits, or that the machine runs, may be sufficient. In other cases more complicated and sophisticated methods must be used. What the tests are, how they will be carried out, and by whom are all important. Standard test specifications for various materials and items of equipment and machinery are given by some of the professional societies and manufacturers' associations and these can be referred to by their appropriate number designation. Tests can be made at the factory, the job site, or both.

Future maintenance Many manufacturers will guarantee or warrant that defects in material or workmanship will be corrected by them for a given period of time. and the terms of such a warranty should be specified. Ordinary maintenance of equipment can be performed by the owner, but in many cases it is more satisfactory if the supplier performs this service. The availability of spare parts and ordinary or special tools can sometimes be a problem and should be anticipated by the specification writer.

EXAMPLE OF PERFORMANCE SPECIFICATION

Sec. 14 Heating and Ventilating

14-1 General Requirements

The printed form of the contract and Section 1 of these specifications are hereby made a part of this specification.

⋮

14-24 Hot Water Tank and Fittings

1. The contractor shall furnish and install a hot water tank of standard, copper, or riveted construction, approximately 2 ½ feet in diameter and 4 feet long, capable of resisting 200 psi internal pressure. It shall be mounted with its long axis horizontal. Minimum clear distance above the floor shall be 1 foot. It shall have a removable heating coil of at least 10 square feet of 16-gage copper tubing, and shall have a capacity to heat at least 150 gallons of water per hour 100°F with steam at zero pounds pressure. The tank shall be complete with all required fittings, including support, and shall be equipped with a suitable dial thermometer, vacuum breaker, and both temperature- and pressure-relief valves.

2. Detail drawings of tank, fittings, and support shall be submitted to the Engineer for approval.

3. Tests. Any tests judged necessary by the Engineer to demonstrate that the tank operates as required by the drawings and specifications will be performed under the Engineer's supervision and in his presence.

6.8 PROPRIETARY PRODUCTS

The third way by which an article or component may be described is by specifying a standard brand or proprietary product. If a product available on the market is known to be satisfactory, specifying by manufacturer's name and model number is the easiest way to describe it. Specifying by this method is widely used for private works, but is usually forbidden in public works. In public works, however, a product may still be described best by designating the manufacturer and his model number, but allowing the contractor to use the product of another manufacturer if it is equal in all respects to the article specified.

CHAPTER 7

Estimating

The purpose of estimating is to calculate the probable quantities and the costs of the many items that comprise a project. Estimates may be made by planners, engineers, architects, owners, contractors, or interested members of the public. They may be made in many different ways and with varying degrees of precision. They may be made before starting a project, during the project, or after completion. They may be for only part of the project or they may be for the entire project. There are many reasons for making an estimate, but in each case the cost of all or a portion of the project is desired.

Estimating is both an art and a science. A good estimator must be able to visualize the completed project from the drawings and other contract documents. He must understand the methods of construction and anticipate all the processes and details of construction that might arise in any project. He must have information concerning the productivity of labor, labor costs, plant and equipment needed and its cost, the cost of all materials required, and the probable cost of all the many overhead items that must be included in the cost of the project. A considerable amount of experience in translating past cost and production into a project that will be performed sometime—perhaps several years—in the future is required.

Unless it is a new different article, the cost of producing an object in a factory or other manufacturing facility can often be predicted with a high degree of accuracy. The costs of the various machine operations necessary can be determined from past records with confidence. Construction costs, however, are affected by so many variable and often unpredictable factors, such as weather, transportation difficulties, or an unstable labor supply, that a construction estimate may present a rather poor picture of the actual cost.

7.1 APPROXIMATE ESTIMATES BEFORE DESIGN

Early in the planning stage of a project some idea of the cost of the project is necessary. The prospective owner of a building, for example, must have some idea of its cost before deciding whether to build or not. If its cost is more than he can finance, or if the cost would exceed its possible financial return or value to him, obviously he should not build it. Before going very far in the design of public works the probable approximate cost must be known in order that plans can be made for financing the venture by appropriations, bond issues, or some other means.

The easiest and least reliable method for determining the approximate cost of a structure is to compare it with a recently completed structure of a similar nature. If a school building in a neighboring town was recently completed at a cost of $1 million and was designed to serve 500 students, one might suppose that next year a school for 1,000 students might be built for $2 million. Such an estimate might be in error by a considerable amount, but by modifying the estimate to take into account changes in labor and material costs and making adjustments for different types of construction, room use, and other factors, an estimate sufficiently accurate for planning purposes might be obtained.

Cost analyses of buildings and other engineering works are constantly being made and their results are regularly printed in technical journals and magazines. The use of these analyses can result in estimates accurate enough for planning purposes. A cost comparison of many hospitals built during the past year, in which their location, size in square feet, number of beds, cost per square foot, type of frame, type of exterior walls, number of stories, and number of elevators was tabulated for each, would be of great help in making an approximate estimate for a planned new hospital.

School building costs are constantly studied and national averages showing low, median, and high costs for different regions of the country are readily available and useful in predesign planning. Cost per pupil or cost per classroom are commonly given, as well as cost per square foot of floor area.

Studies resulting in water or sewage treatment costs for cities of various size can be used to give the approximate cost for a plant based on a per capita cost, gallon-per-day cost, or any other desired parameter.

It should be emphasized that the use of any average cost date is dangerous and must be brought up to date by using appropriate cost indices, geographic factors must be introduced, and adjustments must be made for the type of construction.

7.2 ENGINEERS' OR ARCHITECTS' ESTIMATES

The engineer or architect responsible for the design of a project will have many occasions for determining the cost of the project. Usually the amount of money available is limited and the designer must work within that limit. He will therefore find it necessary to modify the design if it appears that the cost will be too great. Several alternative designs are often studied, and the best solution that fits the budget must be determined. The type of structural system to be used is determined by studying and comparing the costs of several alternatives. Wall- and floor-covering materials and other architectural features must be chosen with the cost as well as the desired esthetic features in mind, so estimating must take place concurrently with the design.

After completion of the design, and before contractors are invited to bid on a project, an estimate should be made by the designer to ensure that the project can be completed with the available funds and also as a check later on the estimates of the bidders. All too often the contractors' bids on a project are greatly in excess of the designer's estimate, and in excess of available funds. In such cases all bids are often rejected, cost-reducing changes in the plans are made, and the entire bidding procedure is repeated. The resulting delay and repeated work are extremely annoying as well as expensive, and certainly indicate a poor estimating procedure by the designer.

In the heavy-construction field there are probably few approximate methods that are at all useful in arriving at the probable cost of a project. The best estimate by the engineer is calculated in exactly the same manner as that used by the bidding contractors. This method will be covered later in the section "Detailed Estimate."

The best estimates of building designers are also made by making the same sort of detailed estimate as that used by the bidding contractors. However, since so many buildings are being built and so many cost data are available, several approximate methods of estimating costs have been devised using these data. Surprisingly good results have been obtained using these methods and, of course, terribly wrong estimates have also been made using the same data. It must be emphasized that the following methods, although often used by engineers and architects with good results, are only approximate methods and should be used accordingly.

7.3 FLOOR-AREA METHOD

This method is commonly used for buildings and depends upon a prior analysis of many buildings of a similar type whose total costs are known. The total cost of the building divided by the square feet of floor area gives a unit cost that can be used to calculate the cost of a similar building.

The question of what floor areas are to be used gives rise to several variations of this method. The most common counts basements, mezzanine, and penthouse areas measured from exterior faces of exterior walls. Covered walkways, porches, and open roofed areas have their areas divided by two.

Another common method recognizes the use made of basement areas as being different from the main floors of the building and uses only one half the unit cost for all basement areas. Roof areas can be included at the same unit cost as the basement.

External features such as parking areas, sidewalks, porches, patios, and extensive or unusual sitework should be treated separately.

7.4 BUILDING-VOLUME METHOD

The great variation possible in ceiling height and, to some extent, roof slope, often makes the floor area method somewhat inaccurate. Another unit cost commonly used is the cost per cubic foot. In using this method the total cost is divided by the volume enclosed from the depth of the footing to the average height of the roof.

As in the floor-area method, many variations are possible. Since the costs of the space between the top floor ceiling and the roof and the basement volumes are not the same as the volume of space on the main floors of the building, these may be treated separately and different unit costs assigned to them.

Both the floor-area and building-volume methods can give good results if used with care and updated by the use of appropriate cost indexes. Unusual architectural features in either the buildings studied or the projected building are difficult to handle by either method and lead to erroneous estimates unless they are treated separately from the rest of the building.

7.5 PARAMETER METHOD

Since it was first described in Alfred P. McNulty's article, "Parameter Estimating Is Fast and Accurate," in *Engineering News-Record*, December 15, 1966, parameter costs of several buildings have been included in each of that magazine's Quarterly Cost Roundup sections. Use of this method depends upon developing cost summaries for large assemblies or systems and relating these costs to the physical measurements or parameters that determine the contribution of that assembly to the total cost of the building. Each assembly is a portion of the building built by a particular building trade.

In using the method, several buildings of a similar nature whose costs are known are analyzed, and the following quantities or measurements are noted:

1. Number of floors, excluding basement.
2. Number of floors, including basement.
3. Basement plan area.
4. Basement area, total.
5. Gross area supported (excluding slabs on grade).
6. Face-brick area.
7. Interior partitions
8. Curtain walls, including glass.
9. Net finished area.
10. Other exterior masonry walls.
11. Number of elevators.
12. Store-front perimeter.
13. Number of plumbing fixtures.
14. Parking area.

For comparison with other buildings, such information as location, dates of construction, type of frame, type of exterior walls, fire rating, typical ceiling height, lobby area, heating and air conditioning requirements, number of rooms, and any unusual architectural features should also be noted.

Cost parameters are then computed for each trade by relating the cost of work done by each trade to the appropriate one of the elements listed above. The cost of structural steel, for example, would be given with respect to number 5, gross area supported; carpentry costs would be related to number 7, interior partitions; waterproofing and damp-proofing with respect to number 10, other exterior masonry walls; acoustical ceiling cost with respect to number 9, net finished area. These unit or parameter costs can then be used to estimate the cost of the building that is being planned or designed.

Several advantages are claimed for the parameter method over the floor-area and building-volume methods. The cost of many features can be calculated as well from preliminary drawings as from finished working drawings. Many items that are necessary for completion of the building do not appear on preliminary drawings, yet their cost would appear in a parameter estimate since they would be part of an assembly or system. Errors and omissions during and after the bidding period, especially by subcontractors, are easier to spot using the parameter method. During the design period the expensive effect of certain design elements can be spotted and corrected.

7.6 COMPLETE ESTIMATE

The total cost of a building or other engineering work will include more costs than those contained in the main contract and the contracts of the subcontractors. A complete or total estimate must include the cost of land, legal fees, architects' and engineers' fees, cost of financing the project, taxes, insurance, costs of permits and licenses, and any other costs in addition to the contractor's costs that the owner is responsible for.

7.7 PROGRESS ESTIMATE

Except for projects of very short duration, the contractor is not expected to finance the project completely, and is paid periodically, usually monthly, by the owner. To determine the amount of these payments, a progress estimate must be made. Frequently this is made by the contractor, but should be approved by the engineer or other owner's representative before payment is made.

In heavy construction in which a unit-cost contract is used, the quantity of each item completed can be measured and the payment due calculated using that quantity and the unit contract price. With a negotiated contract, usually the cost-plus type, the actual costs incurred by the contractor during the previous month plus any agreed-upon amount for overhead should be reimbursed by the owner. The engineer or other owner's representative should, of course, be furnished with receipted bills and other evidence that all costs incurred by the contractor had actually been paid. In the lump-sum type of contract the percentage of the total work of a certain type that has been completed during the previous month is often estimated. That percentage of the amount allocated for that particular type of work is then paid to the contractor.

In most contracts it is stipulated that the contractor will not be paid for all work completed—usually between 85 and 95 per cent—so the amount retained must be deducted from the amount he has earned. There is probably more room for error in the lump-sum type of contract than in the other types, so the owner's representative must be on guard against overly optimistic progress reports by the contractor. This is also a situation in which the contractor might try to charge too much for the first several operations of a project in order to get working capital, thus forcing the owner to finance more of the project than is fair. This is analogous to unbalanced bids, which will be taken up more in detail in Chapter 10.

7.8 FINAL ESTIMATE

When a project under a cost-plus or unit-price contract has been completed, the work must be inspected, total quantities measured, and the amount of the final payment determined. At this time, if all work is satisfactory, all retained amounts must also be paid. The amount of this final payment is therefore determined by the final estimate. The same is true of lump-sum contracts, particularly if there have been numerous changes or extra work done. If there have been no changes or other extras, the final estimate for payment due consists only of the remaining part of the contract price. The final estimate really consists, therefore, of making sure that all contracted work has been completed to the satisfaction of the owner.

7.9 DETAILED ESTIMATE

A detailed estimate, often called a contractor's estimate, is the estimate made by a contractor in determining his bid price. It is, as the name implies, very detailed and is the most accurate type of estimate that can be made. In making this estimate, the quantity and cost of everything that the contractor must supply and do must be calculated.

Making a quantity survey or takeoff is the first step in the estimate. The estimator must examine the drawings and specifications and determine from them the number of cubic yards of excavation, the board-feet of each type and size of lumber, the number of square feet or yards of plaster, the number and type of each size of door and window, the number of cubic yards of each type of concrete, and so on, for the entire project. The number of these items that must be counted and measured is often in the hundreds, so the takeoff man must do a careful and laborious job. Obviously nothing can be omitted and nothing should be counted twice. The quality and completeness of drawings and specifications play a vital role in this part of the estimate. Incomplete work by the designer makes much extra work for the estimator, since he must stop work to ask the designer what is meant by poor drawings and specifications or by omitted details, or else guess at the intent of the designer and allow for a poor guess by a large reserve for contingencies.

After all the quantities have been determined, the price for each item must be found. Complete and accurate cost records are invaluable for this part of the estimate. The cost experience of a contractor in work of a similar nature is the best possible guide in forecasting the cost of an operation in the future. Materials costs will vary somewhat for different locations, depending upon the source of supply and buying conditions. Labor costs incurred in any construction operation will vary considerably between contractors, depending upon the efficiency of the organization and the choice of plant and equipment used for that operation. Thus, although they are available from many sources, national average prices for any operation should not be used by a contractor in preparing a bid, since they cannot accurately reflect the costs for any given contractor. Published prices in trade journals and elsewhere may give the prices charged to the owner for the various construction operations, but since provision for overhead, profit, and contingencies are contained in these prices, they do not really reflect contractor costs. Contractor costs are usually considered as confidential information and are jealously guarded. Divulging cost information to competing contractors would give them an advantage in bidding, so costs and other trade secrets remain the exlusive property of the contractor.

In making a detailed estimate there are five divisions of cost that must always be considered. Not all five are always used in any given operation, but the possibility of their occurrence must be considered. These divisions—labor, materials, plant and equipment, overhead, and profit—will be discussed in the following pages.

7.10 LABOR

Many time-and-motion studies have been made to determine the rate of production of different classes of construction workmen. Thus the number of hours spent by a mason and helper in constructing a brick wall of a certain size can be used to determine the time necessary for a convenient increment, such as 100 or perhaps 1,000 square feet, of wall of a given thickness. Total hours spent by carpenters and helpers in the installation of floor joists or roof rafters or any other carpentry can likewise be determined and expressed in hours per 1,000 board feet or any other desired measure. Working conditions, such as height above ground, weather, and so on, are very important and should be recorded along with the quantitative data. The average results of a large number of time studies can be used reliably to determine the number of hours necessary to perform any desired task. Estimating books frequently contain tables, charts, and graphs which show the results of many such studies, but since a contractor has no assurance that his workmen are of average efficiency, such figures must be used with care.

The composition and size of a team of workmen can affect both the rate and cost of production. The ratio of masons to helpers, for example, or the number of men supervised by the mason foreman can vary considerably from job to job and from company to company, and can have a big effect on production costs. The most efficient size of production team is probably not used as often as it should be, and probably is not even known in many cases. Therefore, the only reliable production figures may be those obtained from company records.

Knowing the rate of production of a given trade, and knowing the total quantity required for the project, the cost of labor can be calculated if the hourly cost of labor is known. Since many construction projects are long term, lasting several years in many cases, the probable cost of labor two or three years in the future cannot always be known with precision. Contracts for most of the building trades are for periods of two or three years. At the end of that time the contract must be renegotiated. Past and present trends can be studied and some idea of the probable cost of labor in the future can be obtained, but very often a considerable error results from this process of extrapolation. The shrewd estimator, however, by his study of the overall construction picture and the national economy, can arrive at a projected cost that is often very accurate.

7.11 MATERIALS

The cost of materials to be used in the near future is usually available from the material supplier. Suppliers are often notified of price changes in the near future by the manufacturer and pass this information along to the contractor. For projects two or more years in the future, however, possible price changes are often a matter of conjecture and the estimator must use judgment in interpreting cost indexes to arrive at the probable cost when the material is to be used on the job. Prices given by suppliers sometimes include transportation to the job and sometimes do not. A clear understanding of the transportation situation is therefore necessary in determining the cost at the job site. Quantity discounts should be obtained whenever possible, and the cost of detail and shop drawings should be included if paid for by the contractor. The cost of inspection, sometimes at the job site and sometimes at the factory, must be allowed for if this expense is to be carried by the contractor. With some materials an allowance for breakage could be used to increase the unit material cost, but the usual custom is to increase the quantity from the material takeoff to include any expected breakage.

7.12 PLANT AND EQUIPMENT

Although it is not possible to draw a fine line to separate plant from equipment, structures and machinery of a permanent (at least not easily moved) nature are usually considered plant, whereas shovels, cranes, trucks, compressors, and other small and easily moved aids to construction are usually thought of as equipment. Not all construction operations will require the use of either plant or equipment, but for those operations that use them, a cost per hour or day or other appropriate time increment is necessary in finding the unit costs for those operations.

Although not all of them always apply to a given operation, three cost items can apply toward the total cost of owning and operating any piece of plant or equipment: fixed costs, transportation costs, and operating costs.

7.13 FIXED COST–RENTAL

Many items of both plant and equipment today are rented or leased by the contractor. There are many situations in which renting is cheaper and more satisfactory than owning. If it is to be used for only a short period of time, it is usually cheaper to rent rather than to buy the equipment and hope to be able to sell it after it is no longer useful. Although it depends upon many conditions, such as location and the financial makeup of the contracting company, there are often tax advantages to the user of rented as opposed to owned equipment.

In the case of rented equipment, the rent must be paid as long as the equipment is in the possession of the contractor, whether he is using it or not. Thus, the rental can certainly be called a fixed cost, and a charge must be made for this rent whenever the machine is used.

7.14 FIXED COST—DEPRECIATION

When equipment is used by its owner, perhaps the most obvious fixed cost is depreciation. Depreciation is the loss in value of anything caused by use, time, or both. As owners of automobiles are sometimes surprised to learn, a new car, whether used or not, will lose several hundred dollars in value during the first year of life, simply because of time. When they are between 2 and 3 years old, most automobiles can be sold for only about one-half the original purchase price—not because they are half worn out, but because of their age. Construction equipment fortunately does not follow the depreciation rules of automobiles, but its value will lessen with time, and this depreciation must be charged to the jobs on which it is used.

Construction equipment should earn money for its owner, and unless the equipment can be kept busy, it is not economical to own it. Since normal wear and tear are proportional to time of use, the age of equipment can be used to determine depreciation. Since equipment is constantly being improved, a machine several years old, even if never used, is not worth as much as the newest similar model of the same equipment, so once again the age of the equipment is the best measure of its loss of value.

The easiest and most commonly used method for calculating depreciation is the straight-line method. This method assumes that the equipment loses value at a constant rate. The annual depreciation, then, is the first cost minus any salvage value divided by the number of years of useful life. The rate of depreciation, expressed as a percentage, is often discussed and equals 100 times the reciprocal of the years of useful life. Although the straight-line method does not give a good picture of the resale value at any age, its use is allowed by the U. S. government for tax purposes, and is very often used. Its ease of application makes it the most popular depreciation method.

The declining-balance method, although somewhat harder to use than the straight-line method, is also acceptable for income tax purposes. The average depreciation rate, determined from the probable useful life, can be doubled or multiplied by 1.5 or some other factor and applied to the book value to determine the depreciation during any year. (Book value is the value during the previous year minus depreciation and represents the value of the equipment as carried on the company books.) The book value at the end of any year is the book value at the beginning of that year minus 1 ½ or 2 times the rate of depreciation times the book value at the start of the year. The minimum book value allowed is the probable salvage value.

The sum-of-the-years-digits method, as well as the declining-balance method, shows a large depreciation in the early life of the equipment and thus shows a realistic book value at any stage in its life. The sum of the years digits is the sum of the numbers or digits in the expected years of useful life. If a piece of equipment has a useful life of 4 years, the sum of the years digits would be 1 + 2 + 3 + 4, or 10. The depreciation during any year would be calculated by multiplying cost minus salvage by the remaining years of life divided by the sum of the years digits.

As equipment wears out or becomes obsolete it must be replaced. Many contractors keep equipment many years beyond its economic life, and many others fail to make allowance for the greatly increased cost of replacement equipment over the cost of their present equipment. One way to make sure that the replacement cost will be available when necessary is to establish a sinking fund. The value of the sinking fund at any time will represent the total depreciation of the equipment to that date. This method, contrary to the declining-balance and sum-of-the-years-digits methods, shows the highest depreciation in the late years of the equipment's life. The sinking fund is set up by depositing the same amount of money at periodic intervals in an investment paying compound interest. Suppose that an amount A

is required after 3 years, the amount deposited each year is P, and the rate of interest compounded annually is r. After 1 year the value of the fund = P (since the deposit is made at the end of the year). At the end of the second year the value of the fund = $P + Pr + P = 2P + Pr$. At the end of the third year the value of the fund = $A = 2P + Pr + (2P + Pr)r + P = 3P + 3Pr + Pr^2$. Multiplying each side of the equation by r and adding 1 to each side gives the following equation:

$$1 + rA = P(1 + 3r + 3r^2 + r^3) = P(1 + r)^3$$

If the annual deposit is desired, it is

$$P = \frac{rA}{(1 + r)^3 - 1}$$

If the sinking fund is set up for n years, the deposit required for an amount A is

$$P = \frac{rA}{(1 + r)^n - 1} \qquad \text{(Eq. 7-1)}$$

With an annual deposit of P, the amount after n years is

$$A = P\,\frac{(1 + r)^n - 1}{r} \qquad \text{(Eq. 7-2)}$$

The deposit, P, can be made annually or any desired number of times per year. If interest is compounded quarterly and deposits are made quarterly, n should be the number of payments rather than the number of years, and r will become the annual rate of interest divided by 4, or the quarterly rate of interest. With these modifications, the two formulas remain the same.

In calculating depreciation in order to determine how much cost should be applied to a construction operation, the straight-line method is usually used. The number of hours of operation that a machine can work in a year is, of course, quite variable, but 2,000 hours is a commonly used number. Thus, if the useful life of a machine is expected to be 6 years, the total depreciation (cost minus salvage) divided by 12,000 will give the hourly depreciation, and it is this figure that should be used in determining the equipment cost for a particular operation.

EXAMPLE OF DEPRECIATION

A piece of equipment costing $56,000 when new is expected to have a salvage value of $14,000 after 6 years of use. Calculate its book value at the end of each year using each of the following:

a. The straight-line method
b. The declining-balance method
c. The sum-of-the-years-digits method

Solution

a. Annual depreciation = 1/6 or (16 2/3%) of cost - salvage
   Amount of annual depreciation = 1/6 × ($56,000 - 14,000) = $7,000
   Value after 1 year = $56,000 - 7,000 = $49,000 (book value)
   Value after 2 years = $49,000 - 7,000 = $42,000 (book value)
   Value after 3 years = $42,000 - 7,000 = $35,000 (book value)
   Value after 4 years = $35,000 - 7,000 = $28,000 (book value)
   Value after 5 years = $28,000 - 7,000 = $21,000 (book value)
   Value after 6 years = $21,000 - 7,000 = $14,000 (book value)
b. Double average depreciation rate = 2 × 1/6 = 1/3 of book value
   Depreciation during 1st year = 1/3 × $56,000 = $18,667
   Book value after 1 year = $56,000 - 18,667 = $37,333
   Depreciation during 2nd year = 1/3 × $37,333 = $12,444
   Book value after 2 years = $37,333 - 12,444 = $24,889
   Depreciation during 3rd year = 1/3 × $24,889 = $8,296
   Book value after 3 years = $24,889 - 8,296 = $16,593
   Depreciation during 4th year = 1/3 × $16,593 = $5,531
   Book value after 4 years = $16,593 − 5,531 = $11,062. (Since book value cannot be less than salvage value, after 4 years it will remain at the salvage value of $14,000.)

c. Total depreciation = \$56,000 – 14,000 = \$42,000
Sum of the years digits = 1 + 2 + 3 + 4 + 5 + 6 = 21
Depreciation during 1st year = 6/21 × \$42,000 = \$12,000
Book value after 1 year = \$56,000 – 12,000 = \$44,000
Depreciation during 2nd year = 5/21 × \$42,000 = \$10,000
Book value after 2 years = \$44,000 – 10,000 = \$34,000
Depreciation during 3rd year = 4/21 × \$42,000 = \$8,000
Book value after 3 years = \$34,000 – 8,000 = \$26,000
Depreciation during 4th year = 3/21 × \$42,000 = \$6,000
Book value after 4 years = \$26,000 – 6,000 = \$20,000
Depreciation during 5th year = 2/21 × \$42,000 = \$4,000
Book value after 5 years = \$20,000 – 4,000 = \$16,000
Depreciation during 6th year = 1/21 × \$42,000 = \$2,000
Book value after 6 years = \$16,000 – 2,000 = \$14,000 (salvage value)

EXAMPLE OF SINKING FUND

A piece of equipment costing \$56,000 when new is expected to have a salvage value of \$14,000 after 6 years of use. In order to replace the equipment after 6 years, a sinking fund is to be established in a bank that pays 5 per cent compound interest.

a. Find the amount of the annual deposit if interest is compounded annually.

b. Find the amount of the quarterly deposit if interest is compounded quarterly.

Solution:

a. After 6 years, the amount of the fund must equal the cost of the equipment minus the salvage or trade-in value, or \$42,000. The annual deposit,

$$P = \frac{rA}{(1+r)^n - 1} = \frac{0.05 \times \$42{,}000}{(1+0.05)^6 - 1} = \$6{,}180 \qquad \text{(Eq. 7-1)}$$

b. The same equation will be used as in part a, but r will now be 1/4 × 5 per cent = 1.25 per cent or 0.0125, and n will be the number of deposits rather than the number of years. Four deposits per year × 6 years gives a value of 24 for n.

$$P = \frac{0.0125 \times \$42{,}000}{(1 + 0.0125)^{24} - 1} = \$1{,}510$$

As expected, the total amount deposited per year is only 4 × \$1,510, or \$6,040, when interest is compounded quarterly, a smaller deposit than when interest is compounded annually and only one deposit is made per year.

7.15 FIXED COST–INVESTMENT COSTS

Besides depreciation, there are other factors which must be considered whether equipment is used or not, which constitute an expense for the owner. Storage (although sometimes treated separately), taxes, insurance, and interest must also be charged against the equipment. Since these are usually based on the value of the equipment, they are usually lumped together and calculated as between 10 and 15 per cent of the average value of the equipment. Interest, at a rate between 6 and 10 per cent, is the largest part of the investment cost, with storage, taxes, and insurance making up the remainder. These costs are variable with location and also with time, so the company's past cost records and predictions for the future are the only reliable means of calculating these costs. Borrowing money from a bank or other source obviously costs money, and this interest is easily calculated and charged against the equipment. Some contractors, however, seem unaware that when equipment is bought outright for cash, that amount is no longer available as an investment and the amount of interest that is thus lost is a legitimate cost and must be charged against the equipment in the same manner as when money is borrowed.

The possible salvage value of equipment is often neglected in calculating the average investment. In this case the average value is one-half the sum of its value at the beginning of its first year and the beginning of its last year of useful life. If A represents its initial cost, and n is the number of years of life, its value at the beginning of the last year is A/n. The average value, then, is

$$\frac{A + A/n}{2} = \frac{A(n + 1)}{2n} \qquad \text{(Eq. 7-3)}$$

This average value, multiplied by between 10 and 15 per cent, will give the annual investment cost. To get the hourly cost, divide by the usually assumed 2,000 hours.

If the probable salvage value can be determined with any degree of accuracy or if it is an appreciable amount, it should be included in calculating the average investment. The annual depreciation will be the initial cost minus the salvage value divided by the number of years of life. The value at the start of the last year will be one year's depreciation plus the salvage value, S. The average value is

$$\frac{A + \{(A - S)/n + S\}}{2} = \frac{A(n + 1) + S(n - 1)}{2n} \qquad \text{(Eq. 7-4)}$$

Unless the salvage value is large and rather certain, however, it is usually not considered. As in the case with no salvage value, an hourly investment cost must be obtained and charged to the equipment.

7.16 TRANSPORTATION COSTS

The installation of conveyor-belt systems, building of necessary buildings, and setting up and dismantling concrete batching or mixing plants can be an expensive and time-consuming operation whose cost must be carefully calculated and appropriately charged to the job.

Some vehicles can be simply driven to the job site, whereas

for some of the larger earth-moving equipment, specially built trailers must be used to haul the equipment to the site. In any case, some expense is involved and must be allowed for.

Some contractors prefer to consider that the expense of moving both plant and equipment to the job site is part of the overhead cost of starting the project, and the transportation charges are thus not thought of as part of the cost of owning or operating the equipment. However it is done, it is a cost that must be included as part of the contractor's price to the owner.

7.17 OPERATING COSTS

Maintenance and repairs Past experience with a particular piece of plant or equipment provides the best indication of probable expense for the future. Manufacturers' literature and trade publications can also be used in estimating maintenance and repair costs. Often a percentage of the depreciation cost is used. The actual depreciation for most plant and equipment is high during the first years of life and becomes lower during later years. Maintenance and repair costs, on the other hand, will be low when the equipment is new and become higher as the equipment becomes older. The number of hours of use expected per year will also decline during its later years because of the increased time spent in repair. The factors will tend to balance each other; therefore, an average maintenance and repair factor can be used. For heavy earth-moving equipment such as tractors and scrapers with a short useful life, between 70 and 100 per cent of depreciation is used, depending upon the type of use the equipment is expected to be subjected to. Large shovels and cranes have a longer life and also a lower factor for maintenance and repairs.

Fuel Electric power, gasoline, and diesel fuel are the most common fuels for all plant and equipment. As for all costs, accurate records of past experience are the most useful and accurate predictors of future costs. Power required for electric motors and the cost of electricity are easily obtained, and the most unknown item in calculating cost is the number of hours per day or year that the motor will be in use.

Gasoline and diesel engines use approximately 0.06 and 0.04 gallon of fuel per horsepower each hour, respectively, at full throttle. It is not expected that equipment will operate for 60 minutes each hour of each day and certainly not always at full throttle. In calculating fuel costs, it is usually assumed that about 50 minutes each hour is a fairly high average of operating time. The time that the engine runs at full throttle will depend upon the type of equipment and upon the way in which it is used. An air compressor might operate at full throttle only during the compression cycle. A bulldozer may operate at full throttle while loading a scraper or moving fill, then move in reverse, still at or near full throttle, to help the next scraper or pick up the next load. A shovel or backhoe while loading trucks would use full throttle only during the excavation part of its cycle. In calculating fuel costs, the actual operating time per hour and the proportion of time spent at full throttle are usually combined into one factor and multiplied by the maximum, or full-throttle, requirement.

Lubrication For some kinds of plant and equipment the cost of lubrication is negligible, or in some cases it may most easily be included in repair and maintenance costs. The cost of lubrication for many vehicles, such as earth-moving equipment, is a substantial item and must be calculated. Lubricating oil, including changes and all labor costs, hydraulic oil, grease, and the many filters in oil and fuel lines must be included. Data are available from manufacturers of equipment pertaining to frequency of oil and filter changes, etc., but these costs are proportional to the fuel costs. A commonly used rule of thumb for excavating equipment is that for diesel power, lubrication costs 50 per cent of fuel costs and 25 per cent of the fuel cost for gasoline engines.

Tires, cables, blades, and other accessories The useful life of many parts of items of equipment is more or less than the equipment itself. For example, tires, when used on excavating machines may last somewhere between 3,000 and 6,000 hours before they must be replaced. Their cost, including between 10 and 20 per cent of cost for repairs, is usually carried as a separate cost item. Such items as cables for cranes and related equipment, dozer blades, and the like, are priced separately from the equipment they are a part of.

Operator The amount of wages paid to the operator of equipment is perhaps the most obvious item of cost and the easiest to predict. In addition to the amount paid to the operator, the employer is responsible for worker fringe benefits and other costs. In the building trades, the amount paid by the employer into the welfare, pension, and vacation fund is about 20 per cent of the base wage rate at the present time. In addition to this amount, approximately 15 per cent more must be allotted to insurance and payroll taxes. Most labor contracts with unions are for a period of two or three years, but beyond the limit of that contract, cost trends must be studied very carefully to determine probable future costs.

EXAMPLE OF OWNING AND OPERATING COSTS

Find the hourly cost of owning and operating a wheeled front-end loader powered by a 150-horsepower diesel engine. The delivered cost of the machine is $40,000 and replacement tires will cost $5,000. Tire life is estimated to be about 4,500 hours. After 7 years the loader's estimated trade-in value will be $6,000. Taxes, interest, insurance, and storage will cost 13 per cent of the average investment. The machine will work about 2,000 hours per year, at full throttle about two-thirds of the time. It will work under rather severe conditions. Diesel fuel will cost 20 cents per gallon.

Solution:

$$\text{Depreciation, less tires} = \frac{\$35{,}000 - 6{,}000}{7\ \text{yr} \times 2{,}000\ \text{hr/yr}} = \$\ 2.07\ \text{per hr}$$

Investment costs:

$$\text{Average investment} = \frac{A(n+1) + S(n-1)}{2n}$$

$$= \frac{\$40{,}000(7 + 1) + \$6{,}000(7 - 1)}{2 \times 7}$$

$$= \$25{,}400$$

$$\frac{13\% \times \$25{,}400}{2{,}000} = 1.65$$

Repairs and maintenance:

90% of depreciation = $0.90 \times \$2.07$ = 1.87

Fuel: $150 \text{ hp} \times 0.04 \times 2/3 \times \$0.20$ = .80

Lubrication: 50% of fuel cost = $0.50 \times \$0.80$ = .40

Tires (including 15% for repairs):

$$\frac{\$5{,}000 \times 1.15}{4{,}500 \text{ hr}} = 1.28$$

Operator: = 8.00

Total hourly cost: = $16.07

This represents the cost to the contractor for the machine plus operator. The cost to be charged to a client should be this figure plus an allowance for overhead and profit.

7.18 OVERHEAD

Overhead is that part of the expense of an operation which does not include the cost of materials, labor, plant, and equipment but does include all the indirect costs. Two kinds of overhead are usually considered—job overhead and general overhead.

Job-overhead items are those which can be charged to a particular job. They consist for the most part of nonproductive personnel, services, and equipment not used directly in the construction. Salaries of engineers, superintendents, clerks, timekeepers, warehousemen, watchmen, and any others not directly connected with production would be classified as job overhead. Costs associated with the project office, such as the actual office building, light, heat, furniture, telephone, water, and sanitary facilities, must be included. Any surveying done in connection with the job, the cost of permits, insurance, and fire and police protection must also be included. Temporary signs, travel, advertising, shop drawings, test boring, photographs, and so on, when applicable to a particular job are classified as job overhead. The total job overhead is sometimes calculated as a percentage of total labor, sometimes as a percentage of the total materials, labor, and equipment cost. Depending upon the particular job, it may go as high as 25 per cent or more of the total labor cost.

General or office overhead refers to the expense of running the company office or headquarters. Salaries of the company officers, managers, estimators, secretarial staff, engineers, and all other personnel not directly involved with any one job must be paid as part of general overhead. The cost or rental of the office itself, the furniture, office and accounting equipment and machines, utilities, parking and storage areas, vehicles such as company cars and small trucks, insurance, advertising, legal and consulting fees, and anything concerned with running the business and not attributable to a particular job are classified as general overhead. The cost of general overhead is reasonably constant unless the size of the company changes greatly, with a large increase or decrease in business, while job-overhead costs usually last only as long as the job. The cost of general overhead is prorated among all the jobs so that each job carries its fair share of general overhead. If a company does $3 million of work per year, the yearly general overhead is $100,000, and the bid price on a certain job is expected to be $600,000, then an overhead cost of ($600,000/3,000,000) × $100,000, or $20,000, would be charged to that job.

7.19 PROFIT

Profit is the amount of money the contractor has left over after paying all the direct and indirect costs in connection with a project. The construction industry is noted as the industry with the highest record of bankruptcies of any in the nation. The possibility of making very little or of losing money is apparently very great, and should be balanced by the possibility of making a high return on the contractor's investment. The amount of profit the contractor should include in his bid depends upon many factors, such as the competition, how badly the contractor wants the job, and the relative difficulty of the job. A contractor might include 2 or 3 per cent profit on a routine building job at a time when competition is high and he wants work badly in order to keep his equipment and work force busy. For a job containing many unknown or dangerous factors such that he runs a high risk, a profit of 20 or 25 per cent may be justified. In general, the higher the risk of losing money, the greater should be the reward.

Since the risk of losing money is so great, unless a contractor can earn considerably more money than a bank or government bonds would pay in interest on his investment, there is little sense in staying in business. The purpose of being in business is to earn money, but including too high a profit in a bid will result in not getting the job, so a large amount of judgment is required in this respect. The statistics of bidding will be studied in Chapter 10.

CHAPTER 8

Quantity Survey - Small Building

A quantity survey or takeoff is undertaken while making a detailed or contractor's estimate. One reason for such a survey is to determine the quantities of all the materials and services required for the completion of the project in order to determine the cost. Another reason for making a quantity survey is to provide a bill of materials so the materials needed for construction can be ordered and delivered in time for their inclusion in the project.

It is not the purpose of this book to provide a complete guide to estimating, but the principles involved in making a quantity survey are simple and this chapter will try to illustrate the procedure for a familiar type of building. Large buildings can be constructed in a large variety of ways, and constant reference to the drawings and specifications is necessary in making a quantity takeoff. There are also many ways of constructing a small, single-family house, but a popular, commonly used method will be used in this illustration to keep references to drawings to a minimum. The principles used in the illustration are few and can be used for any type of construction.

After the quantities of all services and materials have been determined, if the purpose of the estimate is to provide the cost of the building, the unit cost of each item must be multiplied by the estimated quantity. The total of these, then, will be the cost of the building. As in all estimating, the records of past performance by the company provide the best guide to future costs. If no such records are available, such books as *Building Construction Cost Data* by R. S. Means and *National Construction Estimator* edited by Gary Moselle (see the Bibliography) provide nationally averaged costs for building construction. Since these costs are nationwide averages, they should be used with care.

8.1 SITE PREPARATION

Assuming that the house has been designed and that the plans and specifications are complete, the first field operation consists of moving in the required equipment for preparing the site for construction. The site work may consist only of clearing small brush and trees, or it may involve large trees and stumps. In such case, the actual unit cost of the work multiplied by the area, usually expressed in acres, will give the total cost. Sometimes existing buildings, pavements, walls, piping, and so on, must be removed, and of course the cost of such demolition must be known and applied.

Locating the corners of a house on the lot and determining the required elevations can be done by a surveyor or engineer, but is often done by a foreman, superintendent, or in the case of a very small construction company by the contractor himself. Larger firms that might be working on many houses simultaneously, as in the case of a subdivision, would find it to their advantage to retain a surveying crew that did all the required surveying and nothing else.

After locating the corners of the future house, batter boards are set up several feet beyond the limits of the building. Marks and nails at an even number of feet above the cellar floor, base of footing, or other convenient datum will be used as the job progresses for checking floor and other required elevations.

8.2 EXCAVATION

The excavation for a house is usually performed by a backhoe, front-end loader, or sometimes by a bulldozer, depending upon soil conditions and equipment availability. Hand excavation is very expensive and is used only when necessary for precise trimming and cleaning-up operations and to correct errors made by excavating machinery. The length of the house times the width, multiplied by the depth of the cellar, will give the volume of the house excavation in cubic feet. A distance of at least 2 feet beyond the foundation wall should be allowed for working room. Vertical cuts several feet high are unstable and dangerous, so a slope of one or more units vertically to one unit horizontally is often used. The total volume of excavation, expressed in cubic yards, multiplied by the unit cost for excavating will give the cost of the cellar excavation. The type of soil removed, whether easy or hard digging, should be considered in determining the unit cost. The presence of boulders or solid rock that must be blasted will greatly increase the unit cost.

8.3 FOOTINGS

The building to be used in this illustration is a one-story wooden-frame house of platform construction, with concrete foundation walls, plastered interior partitions and walls, and simple pitched roof. Figure 8-1 shows a wall section of the house, and although dimensions have been omitted, much detail of construction can be shown in this single drawing.

The purpose of a footing is to spread the weight of the building over a large area, thus preventing large or uneven settlement. (The same function is performed by skis and snow-shoes.) A thorough foundation analysis for the ordinary house is seldom warranted, and the width of many footings is simply taken as twice the thickness of the foundation wall. The depth is usually equal to the wall thickness. The footing should be placed on undisturbed soil, and the bottom should be below the frost line. In

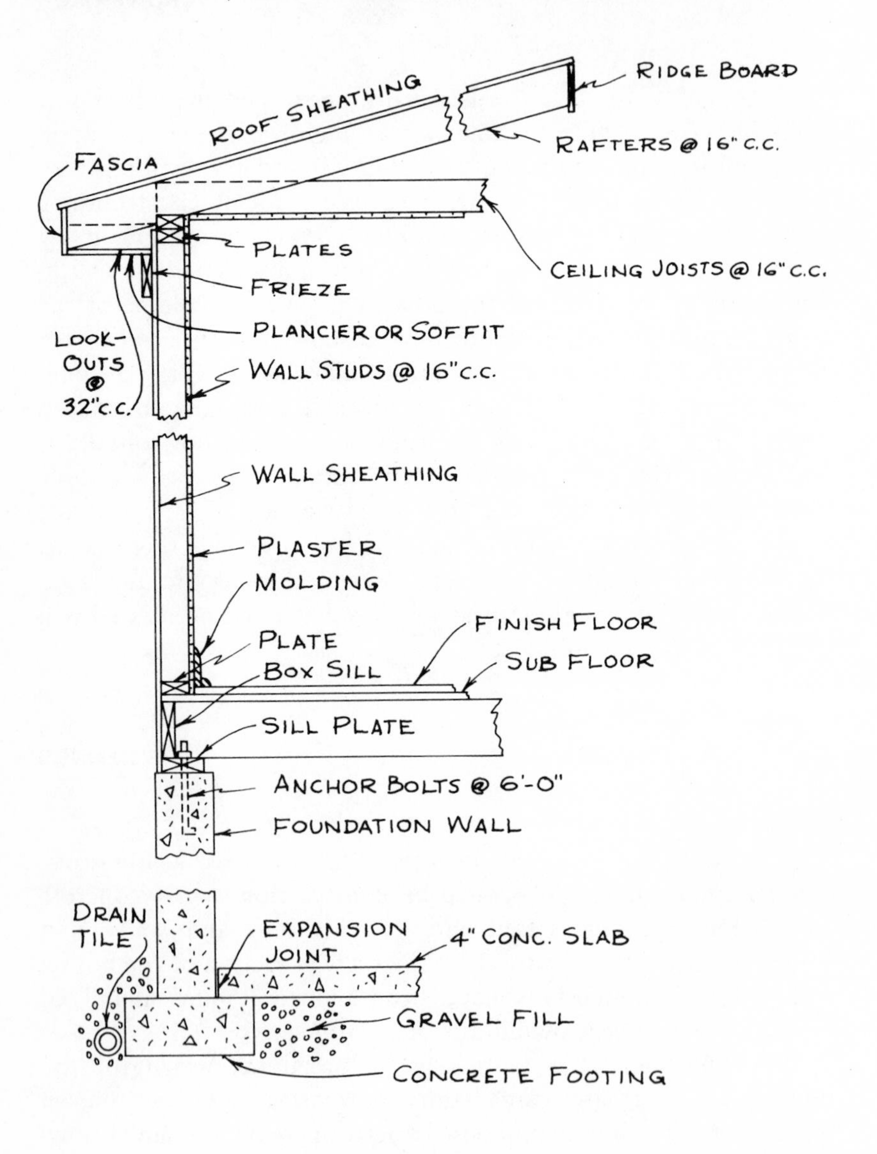

FIGURE 8-1
Typical wall section.

some parts of the country there is no frost, but in the colder regions of the country it may penetrate 5 or 6 feet below the surface. The depth of the water mains in a given locality usually provides a good guide as to the depth of frost.

The perimeter of the house multiplied by the cross-sectional area of the footing in square feet will give the volume of concrete in cubic feet, and this should be expressed in cubic yards. Individual footings under columns, posts, porches, and fireplaces, and so on, are figured separately. For cost purposes, figures should be available giving the cost of concrete in place. This cost must include the cost of forms, the labor of placing concrete, and any curing and finishing that must take place. Most concrete suppliers will furnish concrete in quarter-yard increments, so the theoretical requirement should be increased to the nearest quarter yard. Some waste is inevitable, especially since the depth of excavation cannot be controlled exactly. For large buildings reinforcing steel running parallel with the walls is used, but it is seldom necessary for the normal one-family house.

8.4 FOUNDATION WALLS

In past years foundation walls were commonly made of brick, quarry stone, or native rubble. Today, however, concrete seems to be the most commonly used material, with blocks and bricks used somewhat less often. Block and brick walls are usually 8 inches thick, while 6- or 8-inch thicknesses are used for most poured-in-place concrete. Brick or brick-veneer exterior walls in the house would make it necessary to use 10- or 12-inch-thick concrete foundation walls.

In houses in which a basement is desired, a wall height of 7 or 8 feet is used, while the frost line determines the wall height if no basement is to be provided. The wall thickness times its height multiplied by the total wall length of the building will give the volume of concrete required, and should be expressed in cubic yards. Note that if the perimeter of the building is used, a deduction must be made for the overlapping at the corners.

Deductions for small windows may be made if desired, but their omission would make little difference in the volume of concrete. Since the cost of formwork is a substantial part of the cost, the wall surface area, including all openings, often gives a better indication of cost than concrete volume and may be used as an alternative method.

Anchor bolts are used to prevent lateral movement of the building and also to resist the large uplift forces caused by the wind. The bolts are available in 1/2-, 5/8-, and 3/4-inch diameters and in lengths from 8 to 18 inches. Horizontal maximum spacings of 6 or 8 feet are usually used, but at least two bolts are required for each length of sill plate.

8.5 FRAMING

Most of the lumber used in framing the house is nominally 2 inches thick and has nominal widths of 3, 4, 6, 8, 10, or 12 inches. These dimensions represent the size of the lumber as it left the sawmill. Until recently, lumber was sawed, planed, and worked with a wide range of moisture possible, resulting in a large variation in size when finally sold to the builder. Many attempts have been made to standardize sizes and moisture contents within the lumber industry, and in March 1970 most lumber producers agreed on the sizes to be described below. After being sawed, the lumber is dried to a uniform moisture content of 19 per cent or less, or a maximum of 15 per cent if it is to be designated KD (kiln dried) and run through a planing mill. The resulting thickness of a 2-inch board becomes 1 1/2 inches. The width of 3-, 4-, 5-, and 6-inch lumber is reduced 1/2 inch, and 3/4 inch is taken from the larger sizes. A nominal 2 × 4, then, is (1 1/2) × (3 1/2) inches in cross section and a 2 × 10 is actually only (1 1/2) × (9 1/4) inches.

Lumber is priced and specified by the nominal size in board feet, abbreviated "fbm" for feet board measure, or by the thousands of board feet (Mfbm). A board foot is defined as the quantity of lumber in a board 1 inch thick, 12 inches wide, and 1 foot long. To get the number of board feet in lumber of any size, a simple proportion may be used. The cross-sectional area divided by 12, times the length in feet, equals board feet. The fbm in a 16-foot-long 2 × 10 is [(2 × 10)/12] × 16 feet = 26.7.

Lumber is usually available in lengths of 8, 10, 12, 14, 16, 18, and 20 feet. In some sizes, however, the longer lengths are sold at a premium price and their use should be avoided wherever possible. In estimating, stock sizes must be used in determining how much lumber of a given size must be used, and ordered from the lumberyard accordingly. If lumber 7 feet long is desired, 8-foot lengths might be used and 1 foot cut off and probably wasted. A better solution would, of course, be to cut 14-foot lengths in half, with no resulting waste.

1. Sill plates are of either 2 × 4 or 2 × 6 lumber, sometimes doubled, but often used singly, and are usually sealed to the foundation wall by a thin bed of mortar. The main reason for doubling is to cover the joints between adjacent pieces and to allow for lapping at the corners. The plates must be drilled for anchor bolts. The length of foundation wall or the perimeter of the building will give the number of linear feet required, but this must be broken into practical stock lengths.

2. Box sills or sill headers are of the same size as the floor joists and run at right angles to the joists. The dimension of the building at right angles to the span of the joists will give the length of sill header required, and this must be expressed in stock lengths.

3. Floor joists are the beams that support the floor and all loads, such as furniture, equipment, and people, which are carried by the floor. Most houses and apartment buildings have floors designed to support a load of at least 40 lb per square foot of floor surface in addition to their own weights. Members 2 × 6 or deeper are commonly used, with the depth increasing with the length or span

of the joist. 2 × 6's are adequate for a span of about 10 feet, 2 × 8's for 12 feet, 2 × 10's for 16 feet, and 2 × 12's for 20 feet. These spans are somewhat approximate and vary according to the species and grade of lumber used. The usual spacing is 16 inches center to center, so 12/16 or 3/4 of the run will give the number of spaces between adjacent joists. The run is the length of wall or other support on which the ends of the joists are supported. The number of joists then, is 3/4 × run +1, since there will be one more joist than spaces between joists. The total lumber requirement can be found if the length of each joist is known. When joists are made from two or more pieces of lumber nailed or spiked together to form one long beam, several inches of overlap must be allowed for, perhaps 1 foot or more.

Headers and trimmers are used when an opening such as for a stairway or chimney is to be made in a floor. The header is the member at the head or cut end and runs at right angles to the span of the joist. Headers are always doubled. Trimmers run in the same direction as the joists and support the headers as well as help determine the size of the hole in the floor. They, also, are doubled. Both headers and trimmers are the same size as the joists, and estimating their lumber requirement requires only that their lengths be known.

4. When the span of the floor joists becomes large, the depth of the joists and corresponding weight and cost may become uneconomical. In such cases the joists should be supported at midspan or their third points by larger beams called girders. The girders should be made of large sizes of lumber, but usually are made by spiking three or more 2 × 8's or larger members together. They are supported at their ends by resting in pockets in the foundation walls.

5. Girder posts or lally columns support the girders at intervals of 6 or 8 feet and are supported in turn by piers or isolated footings, about 2 feet square and 1 foot deep. The posts are solid 6 × 6's, or may be made by spiking several smaller members together, or may be steel pipe several inches in diameter. Sometimes the pipes are filled with concrete. Estimating the posts once their size and number are known is obvious.

6. The purpose of bridging is to distribute the loads applied to joists laterally to neighboring joists. Of the three kinds of bridging in common use, cross-bridging is the oldest and most common.

Cross-bridging consists of 1 X 3, 1 X 4, or somewhat thicker material extending from the top of one to the bottom of the adjacent joist. A row of this cross or X-bracing is used for joist spans between 8 and 16 feet, and two rows are used for spans greater than 16 feet. The number of linear feet required is 2 1/2 times the length of the run. A load applied to a joist will be distributed to the joists on each side by the compression in the bridging.

Crossed diagonals made of light-gage steel will distribute the load by acting in tension. If these are used, the number required is equal to the number of spaces between joists.

Solid bridging of the same size as the joists is also used sometimes but is not effective unless it is installed carefully and remains a tight fit between the joists. It resists loads by acting in shear, and must change shape from a rectangle to a parallelogram to permit deflection under a loaded joist. The total length required is equal to the total length of space between joists.

Both the tension and compression types of bridging are fastened to the tops of the joists before any flooring is installed, but their bottoms are usually left free until the floor and some of the expected loads have been applied.

7. Although it should not be classified as framing, the next part of the house to be installed is the subfloor. With its completion there will be a floored-over area the size of the house. It is this platform on which the rest of the house can be built that gives this type of construction its name. Plywood 5/8 or 3/4 inch thick is often used today, and in estimating the quantity the number of standard 4- X 8-foot sheets must be calculated.

Tongue and groove, or dressed and matched lumber of nominal 1 inch thickness is also used, although less than formerly. In forming the mating tongue-and-groove surfaces an additional planing operation is necessary, with the resulting width of the exposed face being reduced 7/8 inch from the nominal width. Nominal widths of 4, 5, and 6 inches will cover widths of 3 1/8, 4 1/8, and

5 1/8 inches. In estimating the number of board feet required of nominal 1 × 4, then, the actual area to be covered should be multiplied by the nominal width divided by the actual exposure and several per cent added for waste, or $4/(3\text{-}1/8) \times 1.03 = 1.32$. In this case 3 per cent has been added for waste. The actual floor area multiplied by 1.32 will give the number of board feet of 1 × 4 dressed and matched lumber to be included in the estimate.

8. With the subfloor in place, the wall framing can be fabricated on the resulting platform, then tipped into place and braced. Wall framing is normally constructed of 2 × 4, but sometimes of 2 × 3, lumber. It consists of a single horizontal floor plate, vertical studs 16 inches on center, and double horizontal plates at the top of the wall.

To determine the required lumber for the plates. three times the total length of wall and interior partition is used, and this length must be broken into stock lengths. Since the plates are made of the same size lumber as the studs, the number of board feet can be calculated.

Studs are usually about 8 feet long. With the common spacing of 16 inches, 3/4 of the run will give the number of spaces between studs. Thus the perimeter of the building times the 3/4 factor would give the number of studs required for the exterior walls if there were no openings and if no special treatment at corners were necessary. Likewise, the total length of interior partition times 3/4 would give the number of spaces between studs. These figures must be modified as discussed below.

No matter what type of interior wall covering is to be used, it must be fastened securely along its entire length. A single stud at an exterior corner would not provide for adequate nailing of the interior wall at that point, so additional material must be provided at all corners. A common way of providing for this is to use three studs at each exterior corner—two separated by 2-inch blocking on one side of the house, and one on the adjacent wall. This method as well as an alternative method are shown in Figure 8-2.

Interior walls must have one more stud than spaces between studs. Some additional material must be used at the intersection of an exterior and interior wall, but the blocking to be used is usually available as scrap from some other part of the construc-

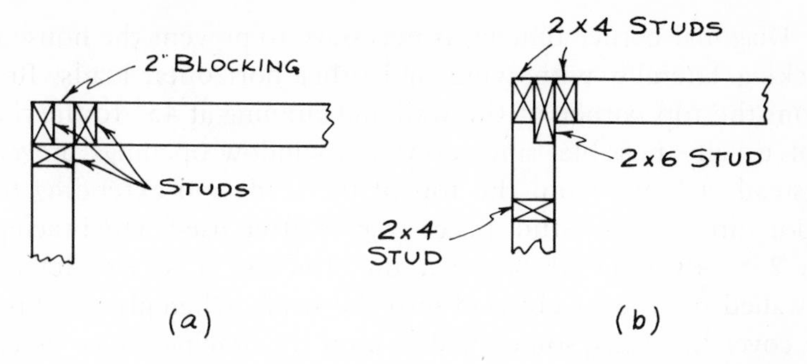

FIGURE 8-2
Studs at exterior corners.

tion. Sometimes it is considered good practice to include a small quantity of miscellaneous lumber of various sizes to be used as bracing, shims, scabs, ledgers, and so on.

Where door or window openings occur in a wall or partition, the studs must be omitted, yet their function of providing support for the roof or floors above must be taken care of. The horizontal members at the top of the opening are doubled 2 × 4's laid on edge for normal-sized openings of 3 feet 6 inches or less and 2 × 6 or larger headers for larger openings These headers are, in turn, supported adequately by doubling the studs or trimmers at each end of the headers. A single 2 × 4 member laid on its side is usually adequate at the bottom of a window opening, although sometimes these members are doubled. For estimating purposes it is not necessary to calculate carefully the exact length of each piece of framing required for the ordinary door or window. Instead, one additional stud for each door and window will care for the headers and doubling of studs required, although many estimators prefer to add two studs per opening.

At gable ends, the length of stud will vary from very little near the eaves to the gable height at the center. The average length will be one-half the gable height, and the number will be 3/4 of the run. The total number of studs will be determined as follows: 3/4 times the perimeter of the building; add 2 for each exterior corner; add 1 for each normal-sized door and window; 3/4 times the total length of interior partition or wall; add 1 for each length of partition or interior wall; add appropriate quantity for gable ends.

9. Diagonal corner bracing is necessary to prevent the house from racking laterally with wind and other horizontal loads. Bracing from the top corner of the wall and running at 45° to the floor is not usually possible, since door and window openings prevent it. Instead, a brace from the top of the studs and extending to the floor three or four stud spaces away is often used. The bracing can be 2 × 4's inserted between the studs, or 1 × 4's "let in" or installed in grooves chiseled into the studs. When plywood is used to cover the walls, some builders omit the bracing. Since bracing is relatively cheap and very effective, its omission may not seem justified.

10. Ceiling joists are sized according to their span, weight of ceiling expected, and possible attic loads anticipated. The presence of partitions running in a direction normal to them will reduce their span. Usual sizes are 2 × 6, 2 × 8, 2 × 10, and 2 × 12. Since they are spaced 16 inches center to center, the quantity required will be 3/4 of the run plus 1. Board-foot measure can be calculated since all rafters will be of the same length.

11. Rafters must be large enough to support the dead load of the roof and its covering in addition to snow load, varying from very little in some states to 35 or 40 lb/ft^2 in some of the colder parts of the country. The usual depths are 6, 8, 10, and 12 inches, and the spacing is usually 16 inches on center. The number required for one side of the roof will be 3/4 of the run plus 1. The length must include the overhang at the eaves and is calculated as the hypotenuse of a right triangle if the rafter span and rise are known.

12. Rafter bracing consists of the ridge board and collar beams, if used. The ridge board should be 1 × 6 or larger, and stock lengths should be ordered. The length of the collar beams can be scaled with sufficient accuracy from the drawings and stock sizes specified. They are positioned one-third of the way down on the rafter from the ridge. Their purpose is to prevent the bottoms of the rafters from spreading outward under the action of

vertical loads. They are not always placed on every pair of rafters, but often on every second or even third pair of rafters.

8.6 ROOF AND WALL SHEATHING

Before plywood became so universally used as a building material, sheathing for walls and roofs was of 1-inch-nominal-thickness boards, sometimes S4S (surfaced four sides) and sometimes D and M (dressed and matched). Some builders prefer to use this rather than plywood because of the extra thickness available for nailing shingles or finish siding. In estimating the number of board feet required for either walls or roof, the actual area to be covered should first be calculated. About 10 square feet per window and 15 square feet for each door should be deducted. The actual areas of oversized doors and windows should be used. About 10 per cent of the gable areas should be added for waste. Since the face widths of nominal 6- and 8-inch D and M are, respectively, 5 1/8 and 6 7/8 inches, and an allowance of about 5 per cent should be included for waste, the resulting areas should be multiplied by (6 / 5-1/8) × 1.05 for 6-inch lumber and (8 / 6-7/8) × 1.05 if 8-inch D and M is used. The result will be in board feet.

If plywood is to be used, the usual minimum thickness is 3/8 inch; 1/2-, 5/8-, and 3/4-inch thicknesses are sometimes used. The area to be covered should be broken down into the number of standard 4- × 8-foot sheets that must be used. In planning the placement of the sheets, it will be noted that at gable ends and around doors and windows there may be much waste, but no overall additional allowance for waste is necessary.

8.7 ROOFING

Saturated felt, an asphalt-impregnated paper, is applied directly to the roof sheathing. Felt is usually priced by the square (100 square feet) or several squares, and is delivered in rolls. The roof area to be covered should be increased by 10 per cent or more to take care of waste and overlapping, and the appropriate number of rolls ordered. Asphalt shingles are also priced by the square, so the number of square feet of roof must be divided by 100 to determine the number of squares. Linear feet of eave, valley, and peak should be treated as roof surface 1 foot wide in determining the total area. The quantity of shingles to be specified should be expressed as the number of bundles required, where one square consists of three or four bundles, depending upon the weight of the shingles.

Wooden shingles or shakes, formerly used almost exclusively as roof covering, are now used mostly when their appearance is an important factor in their choice. They are available in 16-, 18-, and 24-inch lengths, and the number to be used varies with the exposure "to the weather." The first or bottom course is doubled and provisions for ridges and valleys must be made. Tables are available from the manufacturers of shingles showing their coverage for various exposures. A four-bundle square of 16-inch shingles with 5-inch exposure will cover 100 square feet, only 90 square feet if the exposure is 4 1/2 inches, and so on.

8.8 EXTERIOR-FINISH LUMBER

The house by now is nearly completely closed in and protected somewhat from the elements. The roof should be able to shed water and protect the interior from rain or snow, and with the exception of the door and window openings, the walls will provide some protection from the weather. With the exception of the roofing, nearly all of the work thus far can be called rough carpentry. It will be covered as the work progresses and no extremely precise cutting and fitting of the lumber has been necessary. As the building operations continue, more care must be exercised and a higher quality of workmanship must be used. Appearances are very important and better-quality materials must be used in the parts of the house that will be visible after the building is complete. These operations, then, become "finish carpentry."

A typical and simple treatment of the lower edge of the roof at the eave is shown in Figure 8-1. A slightly overhanging roof provides some protection to the walls from precipitation. A larger overhang provides more protection, may provide some shade from the sun, and is often used for decorative purposes. Naturally the larger the overhang the more expensive it becomes, and many houses have recently been built with the minimum possible.

The fascia can be 1 inch or 1 1/4 inches thick. Its purpose is to cover the ends of the rafters, hence its width is governed somewhat by the size of the rafters, but widths of 6 and 8 inches are common. The total length will be the same as the eave of the roof, but about 10 per cent for waste should be included.

The plancier or soffit is also of 1- or 1 1/2-inch lumber, and the width is determined by the required overhang. Plywood should be used for large overhangs, rather than trying to use two or more boards. The length as determined from the drawings should be increased about 10 per cent for waste. Screening and other devices are sometimes installed in the soffit to provide attic ventilation. If a narrow soffit is used, the ends of the rafters can be cut so as to provide horizontal nailing surfaces. For wide soffits a nailing surface must be provided. Short pieces of 2 × 4 nailed to the rafters and the studs, called "lookouts," provide the required surface. The lookouts are spaced 32 inches or more on center. If short enough, they may be made from scrap lumber, but they must be listed in the quantity survey if numerous and long enough to justify it.

The simplest treatment for the joint between the wall and the soffit is a frieze several inches wide and made from 1- or 1 1/4-inch lumber. A more decorative effect may be obtained by using bed mold in place of or in addition to the frieze. As with other finish lumber, the actual length should be increased to allow for waste.

The roof overhang at gable ends is usually less than that at the eaves and often requires only bed mold and a frieze. As at the eaves, however, a fascia and soffit nailed to lookouts may sometimes be used. Estimating the quantities involves only actual lengths increased for waste.

The exterior corners of a house are often dressed up by the use of corner boards. These may be 1- or 1 1/4-inch-thick lumber and several inches wide. At each corner two widths are used to retain the same exposure on adjacent sides of the house. Thus, a 2-inch and a 3-inch width could be used on adjacent sides at the corner, or a 4- and a 5-inch combination. The sizes should be shown on the drawings, and stock lengths with no allowance for waste should be specified.

The finish siding on the exterior walls may extend all the way to the top of the foundation walls, or a more decorative effect may be obtained by the use of a water table and skirting. Many possible schemes may be used here, but the actual length increased for waste is used in making the quantity estimate.

8.9 SIDING

Before installing any siding, the exterior walls should be covered by some kind of building paper. A rosin-sized paper, relatively inexpensive and available in rolls of 500 or more square feet, is often used under many of the wooden varieties of finish siding. The entire outside wall surface must be covered, and special care must be taken around door and window openings to prevent passage of wind, dust, and water.

There are many cross-sectional shapes used for finish siding, but bevel siding has been, and continues to be, the most popular. This type of siding is installed from the bottom of the wall first, and succeeding courses are overlapped in a manner similar to that used for roof shingles. The quantity of siding used depends upon the size of the pieces and the desired exposure to the weather. Bevel siding is usually available in thicknesses of 1/2 and 3/4 inch and widths of 4, 5, 6, 8, 10, and 12 inches, with the narrower widths made of the thinner material. In computing board-foot measure, a 1-inch thickness is assumed. In estimating the siding required, the actual area should be increased about 5 per cent for waste (10 per cent at gable ends), and an allowance should be made for overlapping. If 1/2- × 6-inch siding with a 4 1/2-inch exposed surface is used, for example, the actual wall area to be covered should be multiplied by $(6 / 4\text{-}1/2) \times 1.05$ to give the number of board feet required.

Recently there has been an increased use of plywood for finish siding, and many patterns and wood species are on the market. Some are meant to be installed vertically and some look best horizontally. Various thicknesses are available and they may be installed over sheathing or, if the thicker panels are used, fastened directly to the studs. As in sheathing, the number of standard size sheets should be determined and no further allowance for waste need be made.

8.10 INSULATION

Properly installed insulation is very effective in increasing the comfort of the inhabitants of a house by keeping warm air in and cold air out in the winter, and also by restricting the infiltration of hot outside air into the building during the summer. The control of humidity is also important, both for comfort and preservation of the building.

By far, the greatest amount of heat loss in a building is through the roof, and the greatest amount of insulation is thus placed to prevent this. The heat loss through walls is much less, but wall insulation can do much to improve the comfort level within the house. The air temperature is usually maintained around 70 or 75°F. Cold walls, floors, and ceilings will cause the loss of heat to these surfaces by radiation, making a person feel cold even though the air temperature may be in the proper range.

The passage of moisture through walls and ceiling areas must be prevented. If moisture is allowed to pass into outside walls, condensation will take place at the cold outside wall surface. This condensation can drip down onto the sills and contribute to their damage by rot. Heated moist air that is allowed to escape into the attic space will condense onto the cold roof surface and cause damage to roof sheathing and rafters. Ventilation of attic and wall areas sometimes helps, but the best procedure is to prevent the passage of moisture by proper vapor barriers on the inside surfaces of walls and ceilings.

Vapor barriers may be incorporated into the insulation or they may be separate. Polyethylene film, available in a wide assortment of widths, is often stapled or nailed to both wall studs and ceiling joists to act as a vapor barrier. Thin reflective material bonded to heavy paper, in addition to its action as a vapor barrier, also prevents heat loss through the wall or ceiling by radiation. Since much of the heat loss through an uninsulated wall is by radiation, this type of insulation can be very efficient if properly installed.

Except for the reflective type, all insulation makes use of the fact that still air is an excellent barrier to the passage of heat. Insulation may be made from vegetable or mineral materials and contains dead air spaces within their fibers or cells. These spaces are often microscopic in size, so small that any movement of air within them is impossible.

Insulation in the floor of an attic space is often either poured or blown into place. The material is in a loose form and may consist of mineral or glass wool, expanded mica, or similar material. The thickness of the filled material may be dictated somewhat by the type of heating to be used in the building. The more expensive the fuel, the thicker should be the layer of insulation. The entire space between ceiling joists is often filled, with the resulting insulation being 6 or more inches thick. Loose insulation is usually sold by the bag, with the number of square feet of coverage at a given depth specified. The total number of bags required can be determined from the ceiling area and required depth.

If the insulating material is fastened to a waterproof paper or held between two sheets of paper, the result is a quilt or blanket and is easy to install between ceiling joists or wall studs. Perhaps the most convenient form for walls is an insulating "batt." Batts are available in 16- and 24-inch and other widths and thicknesses, ranging from 1 up to 6 inches. They usually have nailing flanges on each side so they can be fastened to the faces of the studs. They may have built-in vapor barriers or may consist only of heavy paper combined with insulating materials. They are sold by the bag, and the total number of bags can be determined from the wall area to be covered. The thickness used can be the same as the nominal stud size, or it may be an inch or more less. Although the use of quilts or batts is ideally suited to the insulation of walls, they may also be used for ceiling insulation. Loose or fill insulation may also be used for walls, but it must be blown in after the wall structure is complete, hence is not convenient for new construction. Some types of loose insulation may be compressed somewhat with age, leaving a small space containing no insulation at the top of the wall cavity.

8.11 LATH AND PLASTER

Paneling of many sorts, either of wood, plywood, or a wide assortment of plastic materials may be used as interior-wall finish, but the most common for homes and other small buildings is lath and plaster. Drywall construction, in which panels consisting of a gypsum core and heavy paper on each face are nailed to the studs, is also used for house construction. Drywall construction is faster and cheaper than lathing and plastering, and if the joints and nails are properly covered, a good surface that can be painted or papered can be obtained.

In lath and plaster construction, lath is first fastened to the studs, and the plaster is then applied in a plastic condition and allowed to set. Lath formerly was made of wooden strips approximately 1/4 inch thick, 1 1/2 or 2 inches wide, and nailed to the studs with variable, but approximately 1/2-inch spaces between the strips. Today, however, metal or wire lath has supplanted the wooden lath. Flat expanded metal lath is made by pulling laterally a flat strip of metal with properly spaced longitudinal strips. The result is a mesh with rectangular or diamond-shaped openings.

Plaster grounds are small strips of wood of the same thickness as the plaster and 3/4 or 7/8 inch wide. They are usually applied near the ceiling and one or two rows near the floor. One or two rows are also fastened to the studs around all door and window openings. They are used by the plasterer as guides in achieving a smooth finished surface of constant thickness and also provide nailing surfaces for door and window trim, baseboards, and other decorative trim.

Plaster is applied in at least two and usually three coats or layers. The first or scratch coat contains animal or vegetable fibers and is applied to the lath with enough force to squeeze some of the plaster through the openings in the lath to form keys, thus making the plaster adhere firmly to the wall. The fibers help to give these keys strength. Before the surface is completely dried, it is scratched with appropriate tools to make it rough enough for good bond of the next coat. The second or brown coat is applied and brought to a slightly rough, but true, surface and allowed to set. The final or finish coat is thin, but is worked to provide a dense, hard, smooth surface. The total thickness of the three coats of plaster is about 3/4 inch, and a minimum of 5/8 inch is often specified.

For estimating purposes, the area to be covered by lath and plaster, expressed in square yards, is usually adequate. The quantity of raw materials required is left to the discretion of the plasterer, since this work is often done by a subcontractor who specializes in this type of work.

8.12 FINISH WOOD FLOORS

Many finish floor materials are available and in common use today, but oak, maple, and pine tongue-and-groove boards have been used for many years and will probably continue to be for many years in the future. To determine the number of board feet of flooring required, the area to be covered must be multiplied by the nominal board width divided by the actual coverage and about 5 per cent added for waste. Thus, if nominal 1 × 4 boards are used, the actual size will be 3/4 × 3 3/8, and the exposed face will be only 3 1/8 inches. The area to be covered must be multiplied by (4 / 3-1/8) × 1.05 to find the board feet required.

8.13 TILE FLOORS

Floor tiles made from rubber, asphalt, linoleum, and several varieties containing vinyl are used in many rooms, such as kitchen, bath, and special-purpose or utility rooms. Since the subfloor is seldom smooth enough to permit the use of these tiles directly, an underlayment of some sort is usually required. The underlayment can be plywood or hardboard or some other smooth surfaced material, usually available in large 4- × 8-foot sheets. The number of sheets should be calculated from the area to be covered. The tiles come in sizes of 6 × 6, 9 × 9, 12 × 12, and 9 × 18 inches and perhaps 5 per cent should be allowed for waste.

8.14 DOORS AND WINDOWS

Fortunately for the estimator, most doors and windows are specified from standard or stock sizes, and the nomenclature of the many parts need not be learned and referred to. Both doors and windows can be supplied as a complete assembly, as a kit in knocked-down form, or may be custom built. Custom assembly is slow and expensive and is used only when necessary because of unusual size or other features.

Doors and windows are often listed in specifications or on drawings as proprietary products. In making a quantity estimate, the manufacturer's name and catalog or model number must be listed.

If trade names are not used, stock items may still be used. For doors, width, height, thickness, material, number and arrangement of glass panes if used, and any information required to describe paneling should be given. Windows can be described by type, size, material, and size, type, and arrangement of glass lights.

Doors and windows can be supplied with or without trim, and this trim should be noted in the description of the article.

8.15 INTERIOR-FINISH LUMBER

No attempt will be made here to list all the various items of trim and other items of finish that might go into a house. Where their use is indicated on the drawings, the number of linear feet increased by about 10 per cent for waste should appear in the estimate. Information concerning size and location of stairways, cabinets, bookcases, shelves, mantels, closets, and other finish-carpentry items should be given in the plans and can then be counted and estimated.

8.16 COMPLETION OF THE TAKEOFF

Many items remain to be considered before the house is completed, but enough has been covered to indicate how to proceed. Plumbing, heating and ventilating, and electrical work are usually done by subcontract, and their estimating is often left to those who specialize in that type of work.

Omitting items is a common fault of beginning estimators, and to avoid this, a checklist which enumerates all the items that could occur in a building is extremely useful. The use of such a list also provides a methodical listing of the materials in the building, either according to the order in which they will be installed or the trades that will do the installing. Such checklists are often compiled by the experienced estimator in accordance with his personal preferences, or those available in estimating books may be used.

CHAPTER 9

Earth Moving

Nearly all construction involves some earth moving. In the case of repair work, perhaps none or very little earth moving is necessary, but for new buildings some preparation for the foundation must be made, and in heavy construction such as highways, levees, and dams, excavating, transporting, and placing earth constitutes a large proportion of the total work. Water supply and sewerage work involves much pipe laying, and of course trench excavation is part of such activities.

Since the dawn of history men have been changing the shape of the surface of the earth, and the number of men used in some of the excavation accompanying the large structures of antiquity must have been enormous. The use of hand labor except for small trimming and cleaning-up operations is rare today and too expensive for any contractor. The use of machinery has made possible the moving of tremendous quantities of earth for our construction projects at a far lower cost than the use of hand labor would permit. The size and variety of machinery available today for use by the contractor is very large, and a casual examination of any construction magazine will indicate that it is growing even larger. Specialized machines for all possible types of excavation are either on the market or are in the process of being designed. Prices for all excavating machines have increased greatly in recent years, but the unit cost of earth moving remains low owing to the increased productivity possible with the increase in size, speed, and power of the new equipment. Some of the more common machines will be discussed here with the hope that the same methods can be used for more specialized and uncommon equipment.

9.1 BULLDOZER

The most common machine owned or rented by just about all contractors is some kind of a tractor with a dozer blade attached at the front. Several shapes of blade may be used depending upon the type of work, kind of soil, or other job conditions. Straight blades, angling blades, V-blades, U-shaped blades, and land-clearing rakes are in common use. The tractor itself may be on rubber tires or on crawler tracks. Power may be supplied by a gasoline or diesel engine, with the latter preferred for the larger engines. The tractor may be very small, weighing only a few tons, or it may weigh 40 or 50 or more tons. The blade may be cable- or hydraulically controlled.

The dozer is used to push-load scrapers, help maintain haul roads, move earth for short distances, backfill, compact, clear, grub, and strip large areas, and to fell and remove trees and stumps. A ripper, mounted at the back of a track-type tractor, can be used to loosen some soils for easier loading by other equipment, or it may be used to break up pavement for removal. The use of more power and weight has made ripping of some kinds of rock possible, with a lowering of cost for its excavation and removal as compared with the cost if drilling and blasting is necessary.

9.2 FRONT-END LOADER

As its name indicates, a front-end loader consists of a bucket mounted on the front end of a rubber-tired or crawler tractor and is used for handling loose materials. It can be used for light excavating, but its speed and maneuverability makes it best suited for loading trucks and other vehicles and carrying and lifting bulk materials over short distances. The size range of these machines is very large, and new models can be obtained with bucket capacities of 20 cubic yards or more.

9.3 BACKHOE

If a bucket is mounted on the rear of a crawler or rubber-tired vehicle and operated by pulling it toward the vehicle, the result is a backhoe. Pressure on the bucket can be exerted either horizontally or vertically, so digging through fairly hard material is possible. A common and economical use for a backhoe is excavation of trenches for pipes, conduits, foundation walls, and so on. It can be used to load trucks or it can stockpile its spoil for later backfilling. It excavates below its own level and is used for digging basements and other pits, such as for manholes and catchbasins. A very versatile machine often used for trenching and light work around buildings consists of a rubber-tired tractor of about the same size as a farm tractor equipped with both a small bucket on the front and a hoe at the back. Hydraulically operated legs or outriggers at the four corners can be lowered to give the machine stability while digging.

9.4 SCRAPER

A scraper consists of a large bowl mounted on rubber tires and must be towed by a rubber-tired or crawler tractor. The development and increase in use of scrapers since World War II has been rather spectacular, and sizes now range from a bowl capacity of very few cubic yards to monsters used for strip mining with bowl sizes of around 200 cubic yards. Capacities between about 10 and 30 cubic yards seem to be the most common ones used by construction contractors.

Loading a scraper is accomplished by moving it forward as a cutting blade at the forward end of the bowl is lowered into the ground and an apron at the forward end is raised, forming a slot into which a strip of excavated earth is forced into the bowl. When the bowl is filled, the blade is raised, the apron is lowered, and the scraper moves into the dumping area. The load is dumped by lowering the blade to the desired height of fill, raising the apron, and forcing the load out of the resulting slot by an ejector which moves from the rear of the bowl toward the front.

In most types of soil the scraper can be loaded using only its own engine for power, but it is faster and more economical to operate several scrapers on the same excavation and use one dozer-equipped tractor as a pusher. Some scrapers have an engine at the rear to supply more power during both the loading and travel parts of the cycle. Two or more scrapers may also be drawn by the same tractor if hitched in tandem. Elevating scrapers are used in some types of soil for loading without a pusher. The elevator consists of horizontal bars or slats mounted on chains similar to a conveyor belt. As the scraper is drawn forward, the bars dig into the earth and convey it upward and into the bowl.

9.5 CRANE-RELATED EQUIPMENT

Cranes and several types of excavating equipment can be made by varying the attachments on a single power unit. The power unit contains the engine, usually diesel, all the operating controls, and provisions for moving the machine. Controls may be cable- or hydraulically operated; or, for the very large machines, electric controls are used. The unit may be mounted on crawler tracks, may be wheel-mounted with one engine providing power for both moving and operating, or may be truck-mounted with a separate engine for each function.

If a shovel boom and dipper stick are attached to the basic unit, the result is a power shovel. Shovels can be used on any type of soil and are often used in a pit where a bank or face is available to work against. The excavated material may be loaded into trucks or placed on a soil bank. Since the dipper may be crowded into the bank and raised vertically at the same time, the shovel may be used for digging and trimming slopes. Loading into trucks may be at ground level, below, or at the top of a low bank, but the short length of boom prevents loading much above the level of the shovel. Bucket sizes vary from fractional to 10 or more cubic yards, with those between 1 and 5 apparently the most common.

The addition of a hoe boom and dipper to the basic unit provides a backhoe capable of handling nearly any type of soil. When operating, the hoe rests on undisturbed soil and digs ditches, basements, and pits below its own level. The digging action takes place by pulling the dipper toward the excavator. Excavated material may be loaded into trucks or dumped onto a spoil bank for future use. The hoe referred to here, because it is installed on a usually large and powerful basic unit, is capable of very rough and heavy work compared with the hoes referred to earlier. The hoe boom and the shovel boom are both relatively short and strong, making the excavator powerful and precise in action, but useful over only a short radius.

The boom installed on the power unit for use as a crane and several excavating attachments is usually longer and thus weaker than the ones mentioned previously. Lengths of crane booms may be as much as 200 feet. If lifting action only is desired, appropriate hooks, slings, tongs, magnets, concrete buckets, platforms, and many types of boxes may be attached. Counterweights to prevent overturning should be attached according to the working radius and maximum load expected. Large cranes with long booms are commonly used in building construction and have largely replaced the elevators formerly mounted on the outside of a building under construction for the transportation of men and materials to the upper stories.

A clamshell attachment on a crane boom provides a machine used for handling loose materials and excavating below ground level. The clamshell consists of two scoops hinged at their centers and can be opened and closed by cables controlled by the operator. With a long boom, materials can be lifted vertically or horizontally through large distances; thus the clamshell is ideally suited to loading bins and hoppers from stockpiles on the ground. When used as an excavator, material may be loaded into hauling units or stockpiled. The weight of the clamshell limits its thrust into the material; thus it is not well suited to hard excavation.

Like the clamshell, the dragline attachment on a crane boom gives us a machine capable of digging below its own level and depositing the spoil in hauling units or stockpiles. In use, the dragline is lowered to the ground and pulled toward the crane until the bucket is filled. It is then lifted and moved to the dumping area. Since power is applied as the bucket is filled, harder digging can be done by the dragline than can be done by the clamshell. It is often used when earth is removed from a hole or ditch containing water. Neither the clamshell nor the dragline is as precise as hoes and shovels, nor are their rates of production as high. If used to load trucks, a high degree of spillage should be expected unless large-capacity haul units are used.

9.6 ESTIMATING PRODUCTION

As in all estimating, past production records provide the best guide for future work. In many building projects, the excavation will form such a small proportion of the total work that the cost estimate may be completely wrong and still have little effect on the total cost of the building. Very often in building excavation the best machine for the job will not be used for a variety of reasons, with a subsequent high cost. On projects such as dams, highway, and airport construction, earth moving constitutes a large portion of the total project, and care should be exercised in the choice and organization of machinery to do the job. The discussion to follow assumes that the quantity of excavation is significant enough that a thorough job study will be made and an attempt made to secure the best equipment and organization for the job. The principles involved are the same for all equipment, so only two possibilities will be examined for illustrating the method of estimating production and cost.

9.7 SCRAPER PRODUCTION

Three methods are used to estimate excavation costs. The first makes extensive use of the past history of the equipment to be used on a particular job. If similar job conditions, equivalent soil and haul road conditions have been used in the past for the same equipment, these data can be extrapolated to determine production for the job under consideration. Unless nearly identical conditions can be found, however, this method can lead to substantial errors. Another method used with some success on small jobs involves making an estimate of the average haul speed that can be maintained by the equipment. As previously indicated, a large error in this estimate might have little effect on the total cost of a project. If the same method were used in making an estimate for a large-scale project, even a small error in average haul speed could prove disastrous for the contractor. The third method, and the one used here, involves direct calculation. Some judgment must be used in choosing some of the variables, but the use of the method is recommended by several manufacturers of scrapers and other earth-moving equipment.

In calculation production, the time required to move a given quantity of earth must be calculated. Of primary importance, then, are the loading and travel times for a scraper of a certain capacity. The several variable factors that must be considered are as follows:

Material The physical properties of the excavated material must be known. The capacity of the bowl of a scraper is expressed in cubic yards as well as tons. If a material is very light, the volume capacity will govern the load that can be carried, whereas for a heavy material such as iron ore, the weight capacity may be reached before the volume capacity. If neither capacity is exceeded, proper performance of a scraper can be obtained. When soil is excavated it will increase in volume as the spaces or voids between the soil particles are allowed to increase with the decrease in compression of the soil. The percentage of increase in volume is called "swell," or "swell factor," or "per cent of swell." A material with a swell factor of 0.20 or 20 per cent will increase in volume such that 1.20 cubic yards in the scraper will represent only 1.00 cubic yard of bank measure.

If V_L represents loose volume, V_B is bank-measure volume, and s is the swell factor expressed as a decimal,

$$V_L = V_B + V_B s$$

and

$$V_B = \frac{1}{1+s} V_L \qquad \text{(Eq. 9-1)}$$

The coefficient $1/(1 + s)$ is sometimes referred to as the load factor and is listed along with the swell factor for various soils as a convenience in converting from bank to loose measure and vice versa. If the loose capacity of a hauling unit is known, bank-measure capacity equals loose capacity multiplied by the load factor.

Material used for fill is usually compacted or compressed so that it occupies less space than it did in its unexcavated or bank condition. This loss in volume is referred to as "shrinkage." A cubic yard of unexcavated soil with a shrinkage factor of 30 per cent will occupy only 0.70 cubic yard in its compacted state. Both the swell and shrinkage factors can be calculated if the volume of the same weight of soil in bank, loose, and compacted conditions can be measured. Bank yards are usually specified as the basis for payment, but sometimes compacted yards are used.

Haul road The condition of the haul road has much to do with the force required to move a vehicle of any sort. The rolling resistance that must be overcome is a combination of the forces caused by bearing friction, flexing of the tires, and penetration into the road surface. Rolling resistance is proportional to the weight on the wheel and is measured in pounds per ton of weight on the wheel. On a very hard surface such that there is no penetration, the resistance will be about 2 per cent of the weight, or 40 pounds per ton. When penetration occurs, the wheel will always be moving uphill, and the softer the ground, the greater the penetration and corresponding rolling resistance. Rolling resistance on a smooth, slightly flexing roadway will be about 65 lb/ton; with 1 inch or more of penetration it will be 100; about 150 lb for 4 to 6 inches of penetration; and 200-400 lb/ton for soft and rutted roads or in sand. These values can be measured in the field, or intermediate values can be obtained by interpolation.

Grade resistance A vehicle moving up a hill must overcome the force of gravity by providing power, while gravity helps the vehicle overcome other resistance if the vehicle is moving down the hill. The steepness of a slope is usually expressed as percentage of slope or grade and means the number of feet measured vertically for each 100 feet measured horizontally. In Figure 9-1 a vehicle of

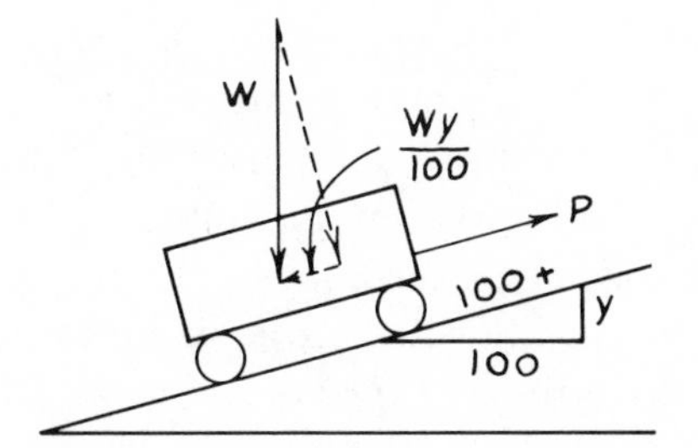

FIGURE 9-1
Grade resistance.

weight W is being pulled by force P up a slope of y per cent. The dashed lines represent the components of W normal and parallel to the slope, respectively. P must be equal in magnitude to the component parallel with the slope, or $P = Wy/100$. The grade resistance to be overcome, then, is always equal to the weight of the vehicle multiplied by 1/100 of the percentage of slope, or 20 lb/ton for each 1 per cent of slope.

Traction The power that can be used depends upon the force of friction between the tires and road surface. This force is equal to a coefficient of friction (which depends upon the roughness of the mating surfaces) multiplied by the weight on the wheel. The coefficient can be determined by experiment with the given road surface and the type of tire to be used. Values for rubber tires vary from about 0.90 on concrete to about 0.10 on ice. The usable power developed by the engine is limited by the force that will cause the tires to slip. Automobile drivers trying to get started on an icy pavement are quite aware that the engine is capable of producing much more power than can be used, and the usable power is limited by the gripping force between the pavement and the tires.

Rimpull The tractive force in pounds between the tires and the road surface is called the rimpull. Empirical formulas can be used to calculate rimpull, but the most reliable figures are those given by manufacturers in their specification sheets for the vehicle under consideration. Specifications for direct-drive machines give both maximum and rated rimpull and maximum speed for each gear, usually in tabular form. Maximum rimpull allows a slight increase in power due to lugging of the engine, but at a reduced speed. This power is used to get the vehicle out of holes or other bad spots. For vehicles with torque converters and power-shift transmissions, the rimpull and speed data are presented in graphical form. For all vehicles the rimpull is highest in the lower gears; maximum speed is attained in the higher gears.

The net towing force that a track-type vehicle can exert on a towed vehicle is designated as drawbar pull and is usually given assuming a rolling resistance for the tractor of 110 lb/ton. Adjustments must be made in the specifications relating speed and drawbar pull available in the various gears if the haul-road characteristics differ from that. Drawbar pull can be measured more easily than rimpull, but manufacturers' specifications are usually relied upon.

Specification sheets furnish power output for engines at or near sea level. At higher elevations the reduced atmosphere causes a reduction in rimpull or drawbar pull available of about 3 per cent for each 1,000 feet of elevation. Some engines are advertised to operate with an excess of air at sea level; thus no reduction of power need be made until the elevation exceeds about 3,000 feet. Engines with superchargers will develop full power even at high elevations. No reduction in speed is necessary when operating at a high elevation since speed is determined by the gear ratios and the setting of the governor.

Fixed time The part of a scraper that is nearly constant for all haul-road conditions is referred to as fixed time and consists of three parts, with a total between 2 and 3 minutes.

Loading time in excess of 1 minute is generally uneconomical and should be avoided if possible by loading downhill with the aid of a pusher. A full load is usually acquired between 3/4 and 1 1/2 minutes, depending upon the size of the scraper and the type of soil.

A round trip from the loading to the dumping area involves at least two turns, and these can seldom be made at full throttle. The actual dumping also takes time, and an allowance of 1/2 minute is usually adequate for both the turning and the dumping.

The higher the top speed expected, the longer the allowance for accelerating must be. Times between 1/2 and 1 1/2 minutes for 3rd and 5th gears hauls, respectively, must be allowed for achieving top speed in those gears. Acceleration also requires additional power, and an allowance of 20 lb. per ton rimpull is sufficient to provide this power.

EXAMPLE OF SCRAPER PRODUCTION

Calculate the probable production of a scraper using the following data:

Haul road—1,800 ft level
2,400 ft up an 8% slope
900 ft down a 6% slope
Return on same road; maximum safe speed 25 mph
Rolling resistance 100 lb/ton, friction coefficient 0.6

Excavated material—dry clay and gravel, 2,700 lb/yd^3 bank measure, 25% swell, 80% load factor

Scraper—two-wheel rubber-tired tractor with 350-hp diesel engine towing scraper with 30-yd^3 heaped capacity

The empty weights on tractor and scraper axles are, respectively, 40,000 and 26,000 lb. Rated load capacity is 70,000 lb. Speed and rimpull data from manufacturer's specifications are given in Table 9-1.

TABLE 9-1

Gear	*Max. mph*	*Rimpull (lb)*
1st	3.00	36,000
2nd	5.50	19,000
3rd	11.05	11,400
4th	19.35	7,000
5th	27.60	5,300

Solution:

Empty weight of unit = 40,000 + 26,000 lb = 66,000 lb = 33 tons

Volume of load = 30 yd^3 × 80% = 24 yd^3, bank measure

$$\text{Weight of load} = 24 \times \frac{2{,}700}{2{,}000} = 32.4 \text{ tons}$$

Loaded weight = 33 + 32.4 = 65.4 tons

Loaded weight on tractor axle = 40,000 + 1/2 × 32,400 = 62,400 lb

Maximum usable power = 0.6 × 62,400 = 37,400

Since this is greater than the maximum rimpull, no slipping of wheels will take place.

Cycle time

1,800 ft, 0% slope, loaded—65.4 tons

Rolling resistance 65.4 tons × 100 lb/ton = 6,540 lb

Acceleration reserve 65.4 × 20 = 1,308 lb

or 65.4 tons (100 + 20) = rimpull required = 7,548 lb

Use 3rd gear (11,400 lb rimpull) at 11.05 mph

$$\text{Time} = \frac{1800 \times 60}{11.05 \times 5{,}280} = \frac{1{,}800}{11.05 \times 88} = 1.85 \text{ min}$$

2,400 ft, 8% adverse slope, loaded

Rimpull required = 65.4(100 + 20 + 8 × 20) = 18,200 lb

Use 2nd gear at 5.50 mph

$$\text{Time} = \frac{2{,}400}{5.50 \times 88} = 4.97 \text{ min}$$

900 ft, 6% favorable slope, loaded

Rimpull required = 65.4(100 + 20 − 6 × 20) = 0

Use 5th gear at 25 mph

$$\text{Time} = \frac{900}{25 \times 88} = 0.41 \text{ min}$$

Total travel time, loaded = 1.85 + 4.96 + 0.41 = 7.22 min

900 ft, 6% adverse slope, empty—33 tons

Rimpull required = 33(100 + 20 + 6 × 20) = 7,910 lb

Use 3rd gear at 11.05 mph

$$\text{Time} = \frac{900}{11.05 \times 88} = 0.93 \text{ min}$$

2,400 ft, 8% favorable slope, empty

Rimpull required = 33(100 + 20 − 8 × 20) = less than 0

Use 5th gear at 25 mph

$$\text{Time} = \frac{2{,}400}{25 \times 88} = 1.09 \text{ min}$$

1,800 ft, 0% slope, empty

Rimpull required = 33(100 + 20) = 3,960 lb

Use 5th gear at 25 mph

$$\text{Time} = \frac{1{,}800}{25 \times 88} = 0.82 \text{ min}$$

Total travel time, empty = 0.93 + 1.09 + 0.82 = 2.84 min

Fixed time:	loading	1.0 min
	turning	0.5 min
	accelerating	1.5 min
	total	3.0 min

Total cycle time = 7.22 + 2.84 + 3.0 = 13.06 min

$$\text{Trips per hour} = \frac{50}{13.06} = 3.83$$

(Sustained production at more than 50 minutes each hour is very difficult to maintain, and 45- or 50-minute hours are often used.)

Production = 3.83 trips × 24 yd^3 = 92 yd^3 bank measure

The average cycle time for a pusher is about 2 minutes. Scraper cycle time divided by pusher cycle time can be used to determine the number of scrapers per pusher. In this example, if cost per cubic yard is desired, the number of scrapers per pusher must be found, that is, 13.06/2 = 6.53, or 7 scrapers per pusher. The total production and hourly cost of 7 scrapers plus 1 pusher must be used in determining the unit cost of the excavating.

It was determined early in the calculations that no wheel slipping would take place under full load since the maximum usable power was greater than the rimpull output of the engine. When traveling empty, however, 40,000 lb × 0.6 = 24,000 lb is the maximum rimpull that can be used. This will increase the accelerating time required in going from 1st to 2nd gear, since not all the power furnished by the engine in 1st gear can be used. This is reflected in the somewhat large allowance of 1 1/2 minutes for accelerating shown as part of the fixed time.

9.8 POWER-SHOVEL PRODUCTION

In estimating the production for a scraper, the fixed and travel times must be determined to calculate cycle time, and nearly the same is true for a power shovel. The operation of a shovel can be broken into four time-consuming elements, and cycle time is the sum of these. In operating the shovel, the dipper is loaded by forcing it into the bank and raising it, the shovel then rotates about a vertical axis until the dipper is above the haul unit, the load is then dumped, and the shovel rotates back to its loading position. The cycle consists of the loading, travel, dumping, and travel elements. Unlike scrapers, there is no fixed time. Many time-and-motion studies have been made by the Power Crane and Shovel Association for different types of soil and with different sizes of dipper. The results of these studies are available in tabular form and give an ideal hourly bank-measure production. These rates must then be adjusted to reflect the actual job conditions. The following variables must be considered:

Depth of cut The most efficient use of a shovel results when it is operated in a pit and digs against a bank or slope. For each size of dipper there is an optimum depth of cut that will give the best production. If the actual depth of cut is too small, excessive crowding of the dipper into the bank to get a full load with one pass will result in lowering production. If the depth is too great, a shallow cut and excessive vertical travel will also lower production by increasing the time required for a full load. Whether the depth is larger or smaller than optimum, a reduction must be made from the ideal output.

Angle of swing In the time trials used for determining ideal production the angle through which the shovel turned from the loading to the dumping position was 90°. Any angle less than that would increase output by requiring less travel time, and a swing angle greater than 90° would decrease output, so adjustments must be made for the actual angle of swing expected.

Job-management conditions Ideal output tables are based on a 60-minute hour and optimum working conditions and equipment. This ideal is seldom met in practice, so a factor must be introduced to describe the actual working conditions. Job and management conditions are classified as excellent, good, fair, and poor. No precise way of defining these categories exists and judgment must be used in their use.

Job conditions are those at the site over which the contractor has little or no control. Weather, space restrictions at the loading site, variations in type of soil or depth of cut, and so on, are usually classified as job conditions.

Management conditions, on the other hand, can be controlled or changed by the contractor. They usually have to do with the condition of the equipment, morale of the workers, available haul units, supervision, maintenance of haul roads and floor of the pit, and the like.

Haul units The size or capacity of the trucks or other hauling units will have an effect on the unit cost of excavating. A not especially reliable rule of thumb often used states that the haul unit should be so sized that it can be filled with four or five dipperloads of the shovel. A too-large truck will spend too much time at the shovel being loaded, and the cost of this nonproductive time will increase the unit cost of the operation. If the trucks are too large it will also be more difficult to balance the outputs of shovels and trucks. Small trucks are more maneuverable than large ones and can sometimes be positioned for loading with great precision, but their small size also requires more care to avoid spilling. Truck size is often determined by the size available to the contractor, but the best procedure is to analyze several sizes and to use the size that results in the lowest unit cost. The number of trucks to be used should be based on the best production (that is, a 60-minute hour) possible by the shovel, since it will operate at that rate part of the time. Slightly more rather than slightly less than the theoretical number of trucks required usually is economical; and, especially in the smaller sizes, providing a spare unit to be used in case of breakdown is good practice.

EXAMPLE OF SHOVEL-TRUCK PRODUCTION

Calculate the probable production of a shovel and trucks using the following data:

Excavated material–sand and gravel weighing 2,900 lb/yd^3 bank measure, with 12% swell factor
Travel and dumping time–12.00 min for a truck of any size
Shovel–2 1/2 yd^3 capacity, 10.0 ft average depth of cut, 60° angle of swing
Job and management conditions–both considered excellent

TABLE 9-2

Output (Cubic Yard/60 Minute-Hour Bank Measure of Power Shovels)

	Dipper Size								
Class of Material	*3/8*	*1/2*	*3/4*	*1*	*1-1/4*	*1-1/2*	*1-3/4*	*2*	*2-1/2*
Moist loam or light sandy clay	85	115	165	205	250	285	320	355	405
Sand and gravel	80	110	155	200	230	270	300	330	390
Good common earth	70	95	135	175	210	240	270	300	350
Clay, hard and tough	50	75	110	145	180	210	235	265	310
Rock, well blasted	40	60	95	125	155	180	205	230	275
Common, with roots and rocks	30	50	80	105	130	155	180	200	245
Clay, wet and sticky	25	40	70	95	120	145	165	185	230
Rock, poorly blasted	15	25	50	75	95	115	140	160	195

TABLE 9-3

Optimum Depth of Cut (Feet) for Power Shovels

Size dipper (yd^3)	*Light, free-flowing materials, such as loam, sand, gravel*	*Medium materials, such as common earth*	*Harder materials, such as rough and hard, or wet and sticky clay*
3/8	3.8	4.5	6.0
1/2	4.6	4.7	7.0
3/4	5.3	6.8	8.0
1	6.0	7.8	9.0
1-1/4	6.5	8.5	9.8
1-1/2	7.0	9.2	10.7
1-3/4	7.4	9.7	11.5
2	7.8	10.2	12.2
2-1/2	8.4	11.2	13.3

TABLE 9-4

Correction Factors for Depth of Cut and Angle of Swing on Power-Shovel Output

Depth of cut in (% of optimum)	*Angle of Swing, (deg)*						
	45	*60*	*75*	*90*	*120*	*150*	*180*
40	0.93	0.89	0.85	0.80	0.72	0.65	0.59
60	1.10	1.03	0.96	0.91	0.81	0.73	0.66
80	1.22	1.12	1.04	0.98	0.86	0.77	0.69
100	1.26	1.16	1.07	1.00	0.88	0.79	0.71
120	1.20	1.11	1.03	0.97	0.86	0.77	0.70
140	1.12	1.04	0.97	0.91	0.81	0.73	0.66
160	1.03	0.96	0.90	0.85	0.75	0.67	0.62

TABLE 9-5

	Management Conditions			
Job Condition	*Excellent*	*Good*	*Fair*	*Poor*
Excellent	0.84	0.81	0.76	0.70
Good	0.78	0.75	0.71	0.65
Fair	0.72	0.69	0.65	0.60
Poor	0.63	0.61	0.57	0.52

Solution:
From Table 9-2, the ideal output using a 90° angle of swing and a depth of cut (Table 9-3) of 8.4 ft is 390 yd^3 bank measure per 60-minute hour. The actual depth of cut of 10.0 divided by 8.4 is 1.2, or 120 per cent optimum. From Table 9-4 the correction factor for depth of cut and angle of swing is 1.11. From Table 9-5 a correction factor of 0.84 is chosen to indicate that, although job and management conditions are excellent, a working time of 50 min/hr is the most that can be maintained over a long period of time.

$$\begin{aligned}\text{Shovel output} &= 390 \times 1.11 \times 0.84 \\ &= 364 \text{ yd}^3 \text{ bank measure/hr} \\ &= 364/60 \\ &= 6.067 \text{ yd}^3\text{/min}\end{aligned}$$

or 112% × 364 = 408 yd^3 loose measure per 50-minute hour. At 100% efficiency a rate of 390 × 1.11 = 433 yd^3/hr or 433/60 = 7.2 yd^3/min is possible, and even though it cannot be maintained for a long period of time, it should be used to determine the number of haul units to provide.

If 10-yd^3 trucks are considered, the loading time will be 10/7.2 = 1.39 minutes, and the total cycle time will be 12 + 1.39, or 13.39 minutes. In a 50-minute hour each truck will make 50/13.39 trips for an hourly production of 50/13.39 ×

10 = 37.4 yd.^{3}. For a haul unit always to be under the dipper, the number provided must be one more than the ratio of travel to loading times. The number required, then, will be 1 + 12/1.39 = 9.65. If 9 are used, hourly production will be 9 × 37.4 = 336 yd^3. If 10 trucks are used, the maximum shovel capacity can be used with an output of 364 yd^3 . The unit cost rather than output should be used to determine whether to use 9 or 10 trucks. If the hourly cost of a 10-yd^3 truck, including the driver, is $20, the unit cost using 9 trucks is (9 × $20)/336 = 53 1/2 cents. If 10 trucks are used, the unit cost will be (10 × $20)/364 = 55 cents. In this case, a slightly smaller production per hour combined with some lost time by the shovel results in the least cost for a truck of this size.

A larger truck should be more economical for a 2½-yd^3 shovel. A 25-yd^3.-capacity truck, costing $30 per hour, can be loaded in 3.48 minutes, make 3.23 trips per hour, and have an hourly production of 80.8 yd^3. The number required is 1 + 12/3.48 = 4.5. The number required can also be determined by total cycle time divided by loading time, or (12 + 3.48)/3.48 = 4.5. Production will be 323 and 364, respectively, for four and five trucks. Using four trucks the unit cost will be (4 × $30)/323 = 37.1 cents and (5 × $30)/364 = 41.2 cents if five trucks are used.

Shovel production at 100 percent efficiency divided by the actual haul-unit production can also give the number of haul units required. Unless the sizes of the excavating and haul units are well matched, this method can give erroneous results. For 10-yd^3 trucks the actual loading time should be 1.65 minutes, total cycle time will be 13.65 minutes, and 3.66 trips in a working hour should produce 36.6 yd^3. Trucks required = 433/36.6 = 11.85. If 100 per cent efficiency loading time of 1.39 minutes is used and an hourly capacity of 37.4 yd^3 is expected, the haul units required would still be 433/37.4 = 11.6. If 9.65 units are provided as previously determined, there will always be a truck under the dipper and maximum production will result. Using more than 9.65 trucks would not increase production and would result in waiting and higher costs. This method of determining haul units should thus be used with care.

9.9 ESTIMATING WITH OTHER EQUIPMENT

The discussion of production of scrapers and shovel-truck combinations applies to large-scale operations for which a carefully planned excavation and haul cycle can be developed and maintained. A similar analysis can be made of operations in which trenching machines, dozers, and other types of equipment are to be used. Time studies have been made by manufacturers of most types of machinery, and their results are available for use by contractors. Time studies made by the individual contractor will more accurately reflect production expected in the future, since they will be made using the men and equipment that will be used on the actual job being considered.

The procedure for all excavating estimation, whether for hand labor or the most sophisticated equipment, makes use of the same principle. A cycle must be established and maintained, and the time for that cycle must be known either by measurement or by calculation. If the quantity of material that can be handled during one cycle can be determined, the total output will equal that quantity multiplied by the number of cycles per hour, per day, or any other convenient time unit. It is not likely that any large-scale excavation will continue at the same rate for several days or months without some lost time because of equipment breakdown, labor troubles, poor weather, and so on, and some allowance should be made for such lost time. If the cost of owning and operating the equipment is known, the cost for the operation, or unit cost if desired, can then be found.

CHAPTER 10

Competitive Bidding

Much of the work done by a construction company is acquired by competitive bidding, and a good estimating team experienced in the preparation of bids is very important. The decision to bid on a project is based on many factors, such as location, type of construction, availability of labor and supervisory personnel, availability of plant and equipment, quality of plans and specifications, reputation of designer and owner, the amount of work in process, the number of competitors, and many other factors.

10.1 GENERAL PROCEDURE FOR PREPARING BID

After the decision to bid has been made, copies of the contract documents must be obtained. Except for the very smallest projects, more than one set of drawings and specifications will be required. The larger the project, the more sets will be necessary, with four or five or more required for a building in the $1 or $2 million class.

All the subcontractors that the prime contractor might

consider using must be notified and requested to submit bids. Since most of the subs will not have their own set of drawings, one or perhaps two sets should be reserved by the prime contractor for the sub. Working space for them should be set aside also.

The quantity survey or takeoff is very important and is most efficiently performed by a team of individuals well versed in reading and interpreting drawings and specifications. The work done by this team is very detailed and requires the utmost care and concentration. Adequate space and privacy for their smooth operation must be provided. Different parts of the takeoff are usually performed by members of the team who are specialists in their field, but a considerable degree of cooperation and checking between members of the team is necessary. Errors and omissions by even the most experienced estimators happen and constant checking of each other's work is one way to keep such mistakes to a minimum.

All materials and equipment necessary for a project must be purchased, taken from existing stores, or rented. The purchasing department, working closely with the takeoff team, is responsible for knowing the best source of materials and equipment and providing prices, dates of delivery possible, and any other desired information.

The cost accounting department is responsible for maintaining records on past jobs and should be able to predict costs on future jobs. The unit costs supplied by them are used with the quantities supplied by the takeoff men and added to subcontractor costs to determine the total bid price, after allowances for overhead and profit have been determined by management and added to the direct costs. The bid price thus determined is then inserted on the proposal form, any other information requested on the proposal can be supplied, and the bid or proposal signed and submitted as requested.

The entire bidding procedure as outlined seems rather uneventful and orderly, but in practice the entire bidding period, especially toward the end, is often quite frustrating and sometimes even frantic. Every bidder hopes to obtain some special knowledge or get some kind of break from his materials suppliers or subcontractors, and waits until the last minute to enter some of his prices in his bid. Many contractors never submit bids until a few minutes before they are scheduled to be opened, and constantly change and revise prices as they receive last-minute information and price quotations. Some of the items that must be considered and that are not evident are discussed in the next several pages.

10.2 SUBCONTRACTORS

Subcontractors are specialists who perform the tasks that cannot be performed efficiently by the work forces of the prime contractor. In a building, for example, subcontractors might be used for heating, ventilating, and electrical work, painting, plastering, roofing, elevators, structural steel, and so on. On a highway job the prime contractor would do the excavating and heavy earthwork and perhaps the paving, and leave such items as steel bridges, seeding and other planting, fences, guard rails, signs, curbs, lighting, and so on, to subcontractors. Relations between subs and prime contractors are often strained for a variety of reasons, and litigation between them is, unfortunately, all too common.

Many subcontractors, although they may be excellent workmen, are very poor businessmen, and often require help in interpreting the drawings and specifications and preparing their bids. No contractual relationship usually exists between the owner and subcontractors, and no bid security is usually requested of the subs. If a subcontractor makes mistakes and consequently makes too low a bid, he may withdraw it or refuse to work for that amount, and with no bid security he can do this without direct penalty. This, of course, is unfair to the prime contractor, who might have used the low bid in preparing his own bid. It is thus important for the prime contractor's estimators to be aware of what a fair price for the sub's work should be and to warn him if his bid is much too low. Accepting a much-too-low bid by a sub, forcing him to do the work for that price, with the resulting loss of money and possible bankruptcy, certainly is, in the long run, poor policy for any prime contractor.

In preparing his bid the prime contractor uses the lowest bids received from his subs, and if he is awarded the contract, he is morally obligated to use those bids. Unfortunately, however, unethical contractors after being awarded a contract will engage in "bid shopping" and try to get the subs to lower their bids. This practice of forcing unfair competition often results in having the subs submit overly high bids to the prime contractors who engage in bid shopping. The high bid can then be reduced when the prime contractor exerts pressure on the sub to cut his bid. In the long run, bid shopping results in higher, not lower prices and does much to increase the tensions and poor relationships already existing between subs and prime contractors.

The bidding practice for subcontractors is seldom as formal and rigid as that for the prime contract, and this can lead to misunderstandings as to exactly what the sub intends to do for his bid price. Bids for furnishing and installing windows as supplied by two subcontractors might seem to be very much different until it is noted that one price includes the interior trim while the other does not. Such discrepancies must be noted by the estimating department so that all subs are really bidding on the same job. The lack of formality for subbids can also lead to a lack of secrecy. A prime contractor who prefers one sub can inform him of the bids submitted by his competitors and thus make certain that the favored one will be low bidder and be awarded the contract.

To avoid some of the unethical procedures and practices used by both prime and subcontractors, bid depositories are sometimes used, especially in public works. Many variations of the depository idea are possible, but essentially they all provide a definite place where bids of subcontractors are deposited for later delivery to the prime contractors, at a specified time prior to the prime-contract deadline. The sub is expected to make identical bids to all prime contractors, but copies of the bid must also be given to the architect or engineer or owner, or perhaps kept at the depository for later reference if bid shopping is to be prevented. Details of bid prices are not made public, and copies to parties other than the prime contractors should not be opened unless complaints of unethical practice make such action necessary. The use of depositories should also make unnecessary the almost frantic last-minute search for subcontractor bids and bid changes by the prime contractor.

10.3 BIDDING TIME

The length of time available for preparing his bid is very important to the contractor in making his decision to bid on a job. Even with an excellent set of contract documents, the time involved in making an estimate—quantity takeoff, prices for all materials, subcontractor bids, planning the construction—can easily take several weeks for large and complicated projects. The time allowed for even small buildings costing less than $100,000 should be at least two weeks, and buildings in the $1,000,000 category call for at least one month.

If insufficient time is allowed for bidding, the bidders will not have time to get the best possible material and subcontractor prices, and of course the owner will suffer because of the high prices the prime contractor must pay. Time spent by the contractor making a careful analysis of the project to determine the most efficient construction methods results in cost savings to the owner, and the lack of this time will correspondingly increase costs for both the contractor and the owner.

As a convenience to the contractors who are used to all the last-minute changes in material and subcontract prices, most bid openings are not scheduled for Monday mornings, but are usually later in the week and in the early afternoon hours. The day following a holiday is avoided, and a day should be chosen that does not conflict with other and similar bid openings.

10.4 COMPLETION TIME

The completion time expected for a project is often stipulated in the contract, and failure to finish at that time makes the contractor subject to liquidated damages for each day that the owner does not have the use of his facility. The time available for construction is therefore very important to the contractor, and he must make a careful analysis of the project and prepare time schedules of the various operations. If the time allotted is not adequate, the contractor must plan the work differently, perhaps using other or additional equipment, planning on overtime work, or using more than one shift per day. These may cost him more money, and his estimated cost should increase accordingly. In some cases, rather than work overtime or add additional shifts, it may be cheaper to plan on late completion and pay the liquidated damages. The cheapest way out for the contractor is the one he should choose.

Sometimes the contractor is asked to furnish the estimated completion date as part of his proposal. The temptation in this case is for him to supply an earlier date than he can possibly meet. If, at the end of that period, the project is still far from complete, there will certainly be strained relations between contractor and owner and the possibility of charges of fraud levied against the contractor.

10.5 UNBALANCED BIDS

In a unit-price bid, if the bid price for an operation is not equal to the fair cost plus a fair share of overhead and profit, the bid is said to be unbalanced. If the bid price is much too high or much too low, it is unbalanced. Usually the overpricing of some operations is balanced by underpricing others.

Bids are sometimes unbalanced to conceal actual costs. A contractor's true costs are considered to be confidential information, and many bidders unbalance many prices simply to prevent the competing contractors from knowing their true costs, which knowledge would be very useful in bidding subsequent jobs.

Very often the cost of getting a project started is high and much of the contractor's funds are tied up in bringing and setting up equipment at the site. One way of forcing the owner to share this expense is by unbalancing the bid so that the first few operations to be completed are overpriced. By submitting high bid prices on the early operations of a project the contractor will also gain working capital to be used in the later phases of the project, thus reducing the amount he must raise by borrowing.

The amount of a unit-price bid is based on the engineer's estimate of quantities. If a bidder feels that any of the estimated quantities are incorrect, he may submit an unbalanced bid in order to make more profit. Every bidder hopes to find some errors in estimated quantities, not noticed by competitors, so that he can unbalance his bid, be the low bidder, and make a large profit. This hope prompts many contractors to make test borings, dig test pits, and consult local citizens—often with great secrecy—to discover some underground conditions that they can capitalize upon. The various possible types of excavation probably offer the bidder the greatest chance either to make or lose money by an unbalanced bid, and the following example will illustrate this.

Suppose that a contractor is asked to bid on a small trenching job. The total quantity of excavation is 10,000 yd^3, and the owner's engineer estimates that 2,000 yd^3 will be rock excavation, the rest being common earth. If $20 and $5 represent fair prices for rock and earth excavation, respectively, a balanced bid would be as shown in Table 10-1.

TABLE 10-1.

Item	*Quantity (yd^3)*	*Unit Price*	*Amount*
Earth excavation	8,000	$5	$40,000
Rock excavation	2,000	$20	$40,000
		Bid price =	$80,000

As a result of investigations of his own, the contractor may feel that the engineer's estimate is incorrect, and although the total quantity of excavation is correct, he thinks that there will be less easy digging and more rock excavation. He decides to unbalance his bid and overprice the rock excavation to make more profit if his gamble is successful. Owners are reluctant to accept unbalanced bids, so the bidder must adjust his prices so that it is not obvious that he is unbalancing his bid. The total bid amount should remain the same, since it is that amount that determines who the low bidder is. His unbalanced bid might be as shown in Table 10-2.

TABLE 10-2.

Item	*Quantity* (yd^3)	*Unit Price*	*Amount*
Earth excavation	8,000	$ 2.50	$20,000
Rock excavation	2,000	$30	$60,000
		Bid Price =	$80,000

The contractor will be paid for the actual quantity of excavation performed, not the estimated quantities. If the owner's estimates of 8,000 and 2,000 yd^3 are correct, the contractor will receive $80,000 whether he has submitted a balanced or an unbalanced bid. If the contractor is correct in his assumption of more rock than anticipated, his payment might be as shown in Table 10-3.

TABLE 10-3.

Item	*Actual Quantity* (yd^3)	*Unit Price*	*Amount*
Earth excavation	7,000	$ 2.50	$17,500
Rock excavation	3,000	$30	$90,000
		Total payment =	$107,500

Using the balanced bid prices for the same quantities, the payment to the contractor would be as shown in Table 10-4.

TABLE 10-4.

Item	*Actual Quantity* (yd^3)	*Unit Price*	*Amount*
Earth excavation	7,000	$ 5	$35,000
Rock excavation	3,000	$20	$60,000
		Total payment =	$95,000

The result of unbalancing his bid would therefore net the contractor an additional profit of $107,000 - 95,000 = $12,500.

It should not be supposed, however, that an unbalanced bid always leads to additional profit. If the actual quantities for earth and rock are 9,000 and 1,000 yd^3, respectively, the contractor would receive $65,000 using balanced bid prices and only $52,500 if the unbalanced prices are used. In this case the contractor's assumption as to the actual proportion of rock to common earth was incorrect, and his gamble would cause him to lose money.

Unless a bid is greatly unbalanced it is often very difficult to detect since it is not unusual for one contractor's costs for an operation to be higher or lower than another's by 50 per cent or more because of variations in construction technique. Because of the possibility that an unbalanced bid might result in excessive payments to the contractor and result in a hardship to the owner, the owner often reserves the right to reject unbalanced bids. The bidder who submits an unbalanced bid therefore should not be too greedy, and must keep the unbalance small enough that it cannot be detected.

Assigning overhead and profit costs to the items in a unit-price bid is usually accomplished by trial. After the direct costs are determined, the percentage of overhead and profit are decided by management, and the resulting additional price can be apportioned as desired. In the example to follow, the items are arranged chronologically, and the contractor would like to be overpaid for the first few items in order to get working capital for the remainder of the project. Management has decided that about 14 per cent of the total direct cost should be added to cover overhead and profit. After some preliminary calculations he might

TABLE 10-5

Item	*Quantity*	*Direct Unit Cost*	*Direct Amount*	*Add to Unit Cost*	*Bid Unit Price*	*Bid Amount*
A	200	$50	$10,000	$10	$60	$12,000
B	4,000	20	80,000	3	23	92,000
C	30	100	3,000	—	100	3,000
D	1,000	5	5,000	—	5	5,000
E	4	500	2,000	—	500	2,000
			$100,000			$114,000

decide as is shown in Table 10-5. The contractor's actual direct costs are confidential, and only the first two and the last two columns would appear in his bid. Although the bid is somewhat unbalanced, none of the items are priced very much too high, and the slight unbalance would not be noticed by the owner. Even if detected, most owners would not reject a slightly unbalanced bid since this method of financing a job is not considered unethical.

10.6 ALTERNATE DESIGNS

Sometimes the final design of a project cannot be decided upon until the costs of the various possibilities are known. If alternate designs for a building are made using reinforced concrete in one case and steel framing as another possibility, contractors may be requested to submit bids on both alternatives, and the method of construction used would be based on the lowest bid received. This method of design seems to place too great a burden on the contractor, and its extensive use by designers should be avoided.

When a definite amount of money has been allocated or appropriated for a project and the owner wishes to get as much as possible or spend all of the appropriation, another use of alternates is indicated. Contractors may be requested to bid on a project, then submit other bids for the basic project with certain features added or deleted. The purpose of the additions and subtractions is to get all the features desired if possible within the appropriated cost, without having to change and redesign if the bids containing the desired features are too high. For example, an architect may request bids on a building as designed, an alternate bid if air conditioning is added, another price if the paved parking area is eliminated, another if carpeting is substituted for vinyl tile or floors, and so on. The priority of alternates must be specified in advance to prevent the alternates from being used to award the contract to a favored contractor.

10.7 ALLOWANCES

It sometimes happens that certain specialty items, such as lighting fixtures and finish hardware, have not been chosen at the time contractors are preparing their bids. They may be informed to include an allowance of a certain amount of money for these items in their bids. The allowance is usually for materials only, and the contractor must include the required amount for labor, overhead, and profit in his bid. As the work progresses, the contractor may be allowed to purchase the materials himself, they may be supplied by the owner, or they may be furnished by a subcontractor chosen by the owner. In any event, the contractor will be reimbursed only the amount spent by him, not the amount of the allowance.

10.8 STATISTICS OF BIDDING

Many contractors rely on intuition and judgment in determining how much allowance they should add to a job's price for profit. How much they want the job, who the competitors are, and how much they think they can add and still get the contract are considered. In recent years much has been written about the use of statistics in compiling bids, and many contractors carefully analyze statistically the past record of their own compared with their competitors' bids in an effort to be low bidders as often as possible. This subject can be extremely complex, and the discussion to follow will merely introduce the possibilities of the subject.

The higher a contractor's bid, the more money he will make if he gets the job. A very low bidder is likely to get the job, but will make very little, or may even lose money. The best bid is some place between these two extremes, and a statistical study of past bids may help to determine it.

The likelihood that an event will occur is called probability and is expressed as a decimal or sometimes as a percentage. A probability of 1.0, or 100 per cent, indicates sure success or certainty, and a probability of 0.0, or 0 per cent, means that there is no possible chance. In flipping a coin, there is an even chance that it will be heads, or there is a probability of 0.5, or 50 per cent. The weatherman's prediction of a 30 per cent probability of rain is based on the fact that in the past it has rained 30 per cent of the times when atmospheric conditions were the same as those now existing.

Profit on a job is the difference between the price received and the cost, but this can be calculated only for those jobs which have actually been performed. Expected profit is a more useful term and refers to the average profit over a long period of time and must take into account the possibility of getting the job. Expected profit is price minus cost on a particular job, multiplied by the probability of getting the job. The task of the bidder is to maximize expected profit and thus make the most money in the long run.

Statistical studies usually consider the bids of the competing contractors compared with the costs of our own company and how many times certain ratios of their bid to our cost have occurred. From that data the probabilities can be calculated of beating the competitor's bid. The probability times job profit will give expected profit, and it is this that we want to maximize for the greatest future profit. Statistical studies are most useful when the bidding patterns of competing contractors are consistent. Much judgment must be exercised in this respect because estimates made by any contractor bidding on a job are likely to contain errors not repeated in other estimates.

EXAMPLE OF THE USE OF STATISTICS

For simplicity, the record of one company that has bid against us 50 times in the past has been compiled (see Table 10-6), and the number of times that their bid was a given ratio of our cost has been noted in the second column. This represents the frequency of occurrence. Column C, Probability, tells how likely an event is to occur and is calculated by dividing the frequency by the total number of occurrences, 50 in this example. In column D the probabilities are expressed in cumulative form. Column E gives the probability of equaling their bid and is obtained by subtracting the cumulative probability from 1.00.

TABLE 10-6

(A) *Their Bid* / *Our Cost*	*(B)* *Frequency*	*(C)* *Probability*	*(D)* *Cumulative Probability*	*(E)* *Probability of Equaling Their Bid*
0.85	1	0.02	0.02	0.98
0.90	2	0.04	0.06	0.94
0.95	3	0.06	0.12	0.88
1.00	4	0.08	0.20	0.80
1.05	6	0.12	0.32	0.68
1.10	7	0.14	0.46	0.54
1.15	9	0.18	0.64	0.36
1.20	11	0.22	0.86	0.14
1.25	5	0.10	0.96	0.04
1.30	2	0.04	1.00	0.00
	50	1.00		

Considering the first ratio, their bid is only 85 per cent of our cost, and if we were awarded a contract at that ratio, we would lose 15 per cent. If we submitted a bid at that price, our probability of being the low bidder is high, and the probability of equaling their bid is even higher—0.98—or 98 per cent of the time we would equal their bid. Bidding the job at cost—a ratio of 1.00—would indicate that we would make no profit on the jobs we got, and we would equal their bid on only 80 per cent of those bid. If we use the ratio of 1.25 and bid 25 per cent over cost, we would make a large profit on the jobs we got, but from column E it is apparent that we would equal their bid on only 4 per cent of the jobs we bid at that ratio.

Tie bids are seldom encountered since contractors often arbitrarily raise or lower their bid prices slightly to keep away from an even number of dollars, thus reducing the chance for a tie bid. In Table 10-6 the figures in column E refer to the possibility of equaling the competing contractor's bid. To avoid ties, the ratios will be adjusted slightly and a new table prepared showing our probability of not equaling, but being lower than, the competitor's bid. If we bid 85 per cent of our cost, we will equal their bid 98 per cent of the time. If we lower our bid slightly, to a ratio of 0.849, we should have a probability of being low bidder 100 per cent of the time.

In Table 10-7, the ratios are now given in terms of both our costs and our bids. Column B gives the probability of being low bidder rather than equaling our competitor's bid. The expected profit in column C is obtained by multiplying the probabilities in column B by the difference between bid price and actual cost.

It should be our goal to maximize expected profits, and as long as we bid against the same bidder, and as long as both his and our bidding patterns remain unchanged, our largest expected profit of 8.05 per cent will be realized if we bid 14.9 per cent above cost and get 54 per cent of the jobs we bid on.

TABLE 10-7

(A) *Our Bid* *Our Cost*	(B) *Probability of* *Being Low Bidder*	(C) *Expected* *Profit (%)*
0.849	1.00	−15.1
0.899	0.98	−9.9
0.949	0.94	−4.8
0.999	0.88	−0.08
1.049	0.80	3.92
1.099	0.68	6.73
1.149	0.54	8.05
1.199	0.36	7.16
1.249	0.14	3.48
1.299	0.04	1.20
1.349	0.00	0.0

Plotting the results of Table 10-7 will not necessarily lead to a more precise result, but as shown in Figure 10-1, the trend of expected profits and losses is quite evident.

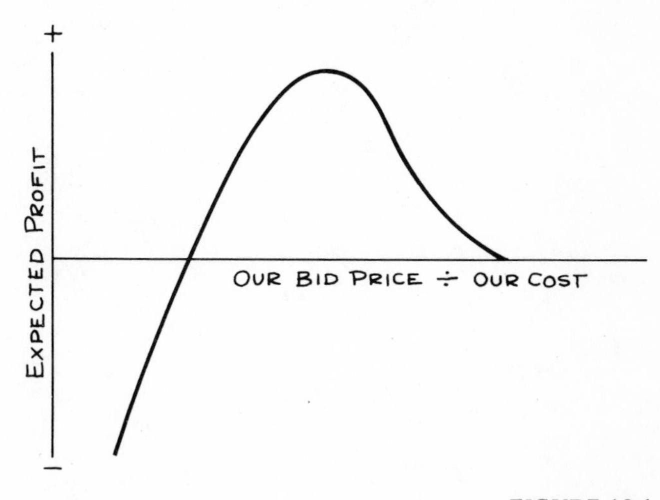

FIGURE 10-1
Variation of expected profit.

If our company usually bids against the same competitors, a table of probabilities of underbidding each of them can be computed in the same manner as that used for only one competitor. The probability of being the low bidder when competing against several is less than that when only one competitor must be beaten, so the probability to be used in calculating the expected profit is the product of the individual probabilities.

If, as often happens, our company bids against many competitors, with the competition being different on different jobs, an average probability of being low bidder must be calculated. On any future job, that probability raised to a power—the power corresponding to the number of bidders—may be used in calculating expected profit and thus determining the best bid.

CHAPTER 11

Critical Path Method

The critical path method (CPM) is a management tool designed to help in the planning, scheduling, and controlling of projects such as those found in the construction industry. It can be used by high-level management or, in more detail, can be used in the field by supervisory personnel. Its use has increased slowly since its inception in the early 1960s, with many users reporting great savings in both time and money by its use. Today many government agencies and some private owners insist on the use of CPM on construction projects, thus forcing its use by many contractors who would otherwise not be interested in using the method.

The development of CPM began in 1956 when men from the DuPont Company's construction division started work with computer experts from Remington Rand on a project designed to find ways to reduce the time required to construct new production facilities. The result of the project, CPM, was first used by DuPont to reduce the shutdown time required for periodic maintenance of some of their production units. Increased uses for the method have been found since then for construction and many diversified industrial projects.

At about the same time that CPM was being developed, the U. S. Navy was concerned about the time required to make some of its missile systems operational, particularly the Polaris submarine. The result of their research was Project Evaluation and Review Technique (PERT). It has been claimed than more than one year of time was saved in the planning and construction of Polaris by the use of PERT.

CPM and PERT are similar in nature, one of the differences being that the time required for performing activities is assumed to be known with CPM but is determined statistically with PERT. The development of Polaris contained many completely unknown operations, and the use of PERT was therefore indicated. Most construction consists of a series of repeated operations using well-known techniques, hence CPM is usually preferred. Both systems were developed for use with computers, but both of them can also be used without computers, since the calculations used are very simple arithmetically. Both methods can be used for both large and small projects, with the use of a computer not economical until the project size exceeds perhaps 50 or 60 individual operations. Computer programs are available for both methods from the manufacturers of computing equipment. Since it is slightly simpler, CPM will be used exclusively in the discussion to follow.

11.1 PLANNING

The first step in planning a project, whether using CPM or not, is to break the project into separate tasks, work units, jobs, operations, or activities. These must be easily defined units that require the use of resources such as labor, materials, or equipment; or they may require only time. Such activities as excavating, hanging doors, and installing plumbing require the use of resources. Curing concrete and waiting for the delivery of materials require time rather than the use of resources and are thus also classified as activities.

The necessary sequence of activities and the relationship between them is an important part of the planning process. In determining them, three questions are often asked. For any particular task:

1. What activities can be performed at the same time?
2. What activities cannot start until completion of this activity?
3. What activities must be completed before this one can start?

The results of the planning can be shown graphically in an arrow diagram. Each activity can be represented by an arrow, with the tail of the arrow representing the beginning and the head the completion. The arrangement of the arrows shows the sequence and the relationships between the activities. The arrows are not vectors, and their lengths and directions are not significant. A circle or node is usually placed at the head end of an arrow and represents a checkpoint in time, or the completion of an activity. In Figure 11-1 the arrow diagram shows that activity *A* is the beginning of the project, and must be completed before *B* and *C* can start, and *B* and *C* can be performed concurrently. *B* must be completed before *D* can start, *C* must be completed before *E* can start, and both *D* and *E* must be completed before *F*, the last activity of the project, can start. The nodes are usually numbered such that their numbers increase in the direction of the arrows. Since an activity may be named by the nodes at head and tail, an attempt is often made to relate node numbers to the account numbers as used by the accounting department. Nonconsecutive numbers are often used to allow the insertion of other activities as the job progresses.

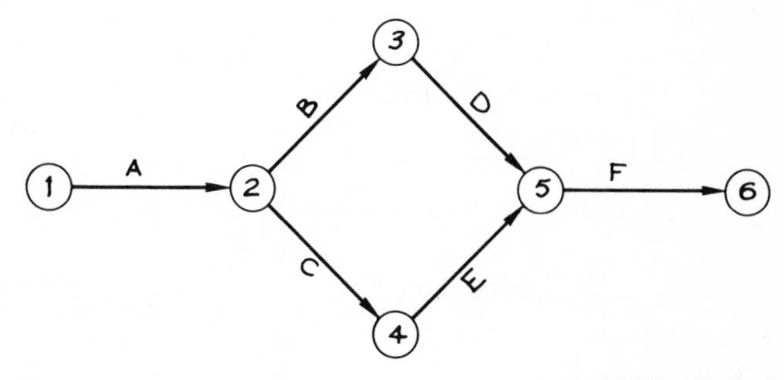

FIGURE 11-1
Arrow diagram.

An activity can be defined as the performance of a specific task which makes use of time or resources or both. Only one exception to this rule exists and it is illustrated in Figure 11-2. If it is determined that activity E cannot start until the completion of B, and the dependence of D upon B, and E upon C still exist, the dashed arrow between nodes 3 and 4 is inserted. Since it requires neither time nor the use of resources, it is not, strictly speaking, an activity, but it shows the relationship between activities and is called a dummy activity, or simply dummy. Had the head of the arrow been at the other end, the start of D only after the completion of C would have been indicated. Another use of a dummy is shown in Figure 11-3. Each activity should have a unique name, as determined by the node numbers, and activity 8-10 is inserted to give M and N different names.

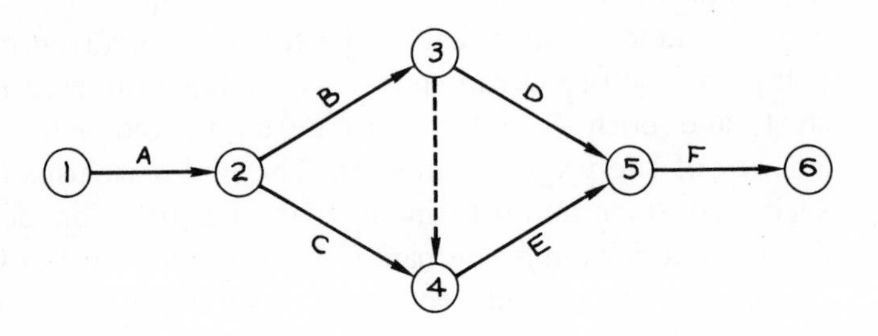

FIGURE 11-2
Arrow diagram with dummy activity.

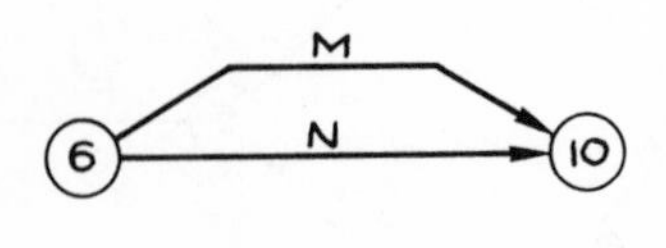

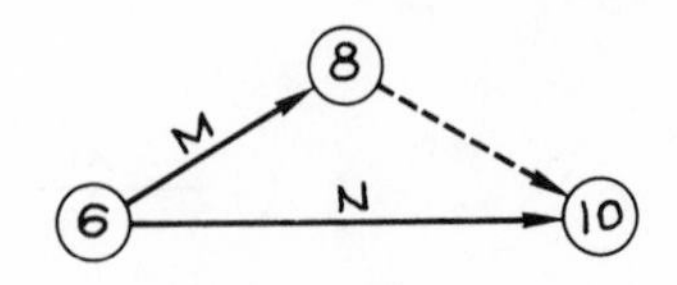

FIGURE 11-3
Node numbering.

11.2 SCHEDULING

The first step in scheduling a project is to determine the time required for each activity. As soon as the method to be used has been determined, company production records can be used to determine the time required, or the duration of each activity. Allowance for weather and other delaying factors should be incorporated for activities, particularly those of long duration, that experience indicates may be subject to delay. The duration of each activity should be calculated using methods and equipment that the contractor actually expects to use, and should not be based on overtime work or equipment not available to him.

After the arrow diagram has been completed and the duration of each activity determined, the calculations leading to the construction schedule can be made. Some of the terms and abbreviations, in the order in which they are used, are defined as follows:

Early or earliest starting time (ES or EST) The earliest time at which an activity can begin. All activities that must be performed before starting this one must be completed.

Duration (Dur. or D) The time required to perform an activity.

Early or earliest finishing time (EFT or EF) The earliest time at which an activity can be finished. ES + D.

Late or latest finishing time (LF or LFT) The latest time at which an activity can be completed without affecting completion of the entire project.

Late or latest starting time (LS or LST) The latest time that an activity can be started without affecting completion of the entire project. LS - D.

Total float (TF) Maximum delay in an activity without affecting the completion time for the entire project. Total float is also called slack time. LF - EF or LS - ES.

Free float (FF) Maximum delay in an activity without delaying the early starting time for a following activity. ES of following activity - EF of this activity.

Critical path The unbroken chain or sequence of activities that will result in the minimum completion time for a project. All activities on the critical path must have zero total float. The CP is also the maximum computed time for all possible paths.

EXAMPLE OF PLANNING AND SCHEDULING

A small construction project has been broken down into seven different activities. Each activity will require different resources, so a resource conflict is impossible. The relationships between activities and the durations of each in days are summarized in Table 11-1.

TABLE 11-1

Activity	*Duration*	*Preceding Activity, Comments*
A	3	None. First activity of project.
B	2	*A*.
C	4	*A*. Can be concurrent with *B*.
D	5	*B* and *C*.
E	3	*C*.
F	7	*C*. Can be concurrent with *E*.
G	2	*D*, *E*, and *F*. Last activity of project.

1. Draw an arrow diagram for the project.
2. Determine the critical path.
3. In tabular form show ES, EF, LS, LF, TF, and FF.

Solution

1. The mechanics of drawing such a simple arrow diagram should give little trouble, but it should be realized that a diagram for 50 or 100 or more activities can lead to much trial-and-error work and much redrawing. A methodical

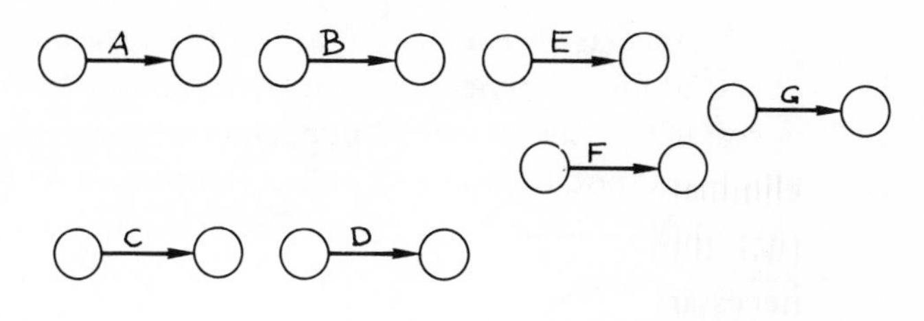

FIGURE 11-4
First step in drawing arrow diagram.

approach to a large project might be to arrange the activities, unconnected, in a somewhat random pattern, as shown in Figure 11-4. The addition of dummy activities to show the relationships between the real activities results in the diagram of Figure 11-5. For calculating purposes the presence of so many dummies does no real harm, but a more concise and understandable diagram results if as many as possible can be

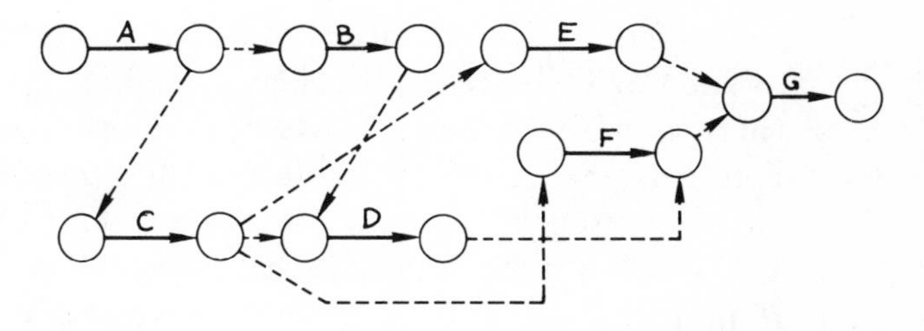

FIGURE 11-5
Arrow diagram with dummy activities.

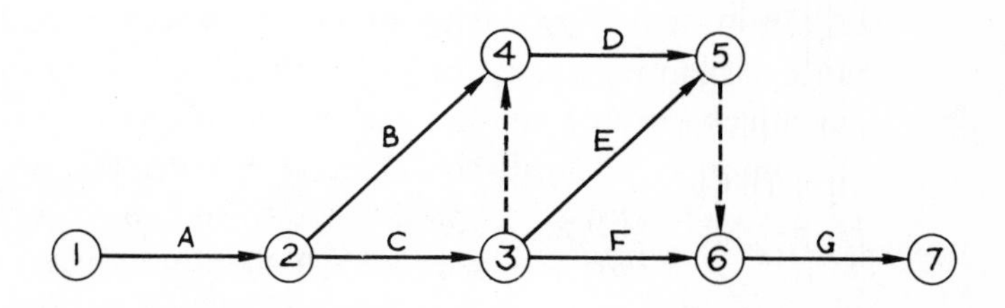

FIGURE 11-6
Completed arrow diagram.

eliminated, as shown in Figure 11-6. It should be noted that two dummy activities are still required. The first, 3-4, is necessary to show that *D* depends upon both *B* and *C*, but *E* and *F* are not related to the completion of *B*. Dummy 5-6 is necessary so that *E* and *F* will have unique names in node terminology.

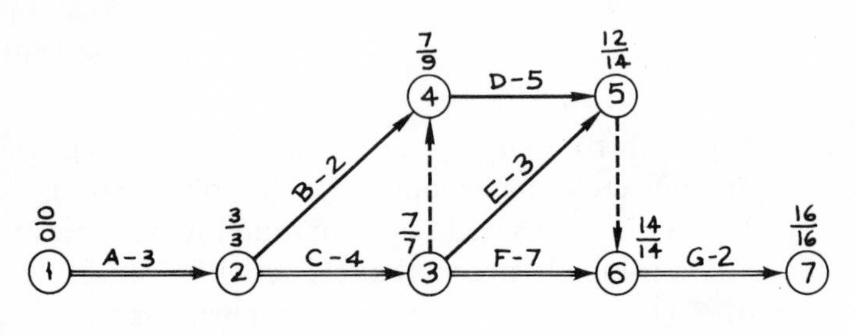

FIGURE 11-7
Arrow diagram with earliest possible and latest possible times.

2. In Figure 11-7 the arrow diagram is repeated and the duration of each activity has been added. At each node, also, what resembles a fraction has been added. The numerator represents the earliest possible time for the event that the node represents. If node 3, for example, represents a checkpoint in time, it means that after 7 days have elapsed, all activities leading into that node have been completed and *D*, *E*, and *F* can be started. The earliest possible time at node 3, then, is the sum of the durations of *A* and *C*. There are two paths leading to node 4. The path from 1 to 2 to 4 requires 5 days, and from 1 to 2 to 3 to 4 requires 7 days. Since *D* cannot start until *all* activities leading into node 4 have been completed, the earliest possible time to record for that node is 7. Similarly, at node 5, path 1-2-3-4-5 requires 12 days and is used in preference to the 10 days required by path 1-2-3-5. In all cases in which more than one time is possible, the *largest* must be used. The critical path is the unbroken sequence of events that results in the largest

completion time for the entire project. In this case it is 1-2-3-6-7. Activities *A*, *C*, *F*, and *G* are critical, and no delays in these activities are possible without delaying completion of the entire project.

The denominator at a node point represents the latest possible time for the event that the particular node represents. These times are calculated by starting at the end of the project and working toward the beginning by subtracting durations. The latest possible time to arrive at node 7 is 16 days, at node 6 it is 16 minus the duration of *G*, or 14 days. The latest possible time to arrive at node 5 is 14 minus the duration of dummy 5-6, or 14 days. For node 4, 14 minus the 5-day duration of 4-5 gives 9 units. At node 3 there are three possibilities. From 5 to 3, 14 minus *E*'s duration of 3 gives 11. From 4 to 3 we get 9 minus zero, or 9. The path from 6 to 3 gives 14 minus 7, or 7 days. If we arrive at node 3 after 11 days, the completion of the project will require 11 plus the duration of *F* and *G*, or 11 + 7 + 2, or 20 days. If we are to complete the project in 16 days, the latest possible time to arrive at node 3 is 7 days. In all such cases in which more than one possibility exists, the *minimum* must be chosen. It should be noted that the critical path goes through those nodes for which the earliest possible and latest possible times are identical, that is, those nodes for which the numerator and denominator are the same. In the diagram the critical-path arrows have double lines to distinguish them from activities which are not so important from a time standpoint.

3. The tabular data requested in the problem is shown in Table 11-2. The earliest starting time for each activity is represented by the numerator at the node (Figure 11-7), representing the start of that activity, and is tabulated under ES. The earliest finishing time, EF, is ES + Dur. The latest finishing time, LF, is the denominator at the node representing the finish of an activity, and LS, the latest starting time, is LF minus the duration of that activity.

TABLE 11-2
Scheduling Data for CPM Example

Activity	*Duration*	*ES*	*EF*	*LS*	*LF*	*TF*	*FF*
A	3	0	3	0	3	0	0
B	2	3	5	7	9	4	2
C	4	3	7	3	7	0	0
D	5	7	12	9	14	2	2
E	3	7	10	11	14	4	4
F	7	7	14	7	14	0	0
G	2	14	16	14	16	0	0

The calculations for total float, TF, contain some checks on arithmetic. The difference between early and late starting times should be the same as the difference between early and late finishing times, and they equal total float. Thus, for activity *B*, total float equals both 7 - 3 and 9 - 5, and 4 is the calculated TF, using either starting or finishing times. Since total float represents time that can be wasted or delay that may be permitted without delaying the finish of the project, all activities on the critical path have zero total float.

Free float has no effect upon the project completion time but is important in scheduling subcontractors. For any activity it is calculated by subtracting the EF of that activity from the ES of the following activity. The FF for E, for example, is ES for G minus EF of activity E, or 14 - 10 = 4.

11.3 COMPRESSING THE TIME SCHEDULE

The activity durations used in CPM calculations are based on the contractor's normal construction methods. Normal practice is based on the contractor's experience and his search for the optimum size of his working crews and best choice of equipment. Each contractor hopes that his method of performing an activity is the best possible and the least expensive. Any changes made in this method should result in increased cost as well as a change in duration of the activity. If a crew of six men plus a foreman is thought to be the most efficient for a certain activity, the addition of two men would probably reduce the duration but would also increase the cost somewhat by lowering the efficiency of each workman. Increasing the number of pieces of equipment can also reduce the duration, but any increase beyond the optimum number will at the same time reduce the unit productivity as costs increase. As shown in Figure 11-8, most activities can be performed in less time than normally, but at an increased cost, by using more men, equipment, and other resources. The minimum possible time required for an activity is its crash time, and the cost associated with it is the crash cost. Beyond that point, the addition of more labor or equipment will result in unnecessary crash, with workmen or equipment getting in each other's way, and with high costs and no increase in production. Reducing the resources normally allotted to an activity will also increase the cost and at the same time will increase the duration, so this practice should be avoided if possible. As shown in Figure 11-8, the relationship between the cost and duration is not linear, but for purposes of compressing the time schedule it is usually considered to be a straight line.

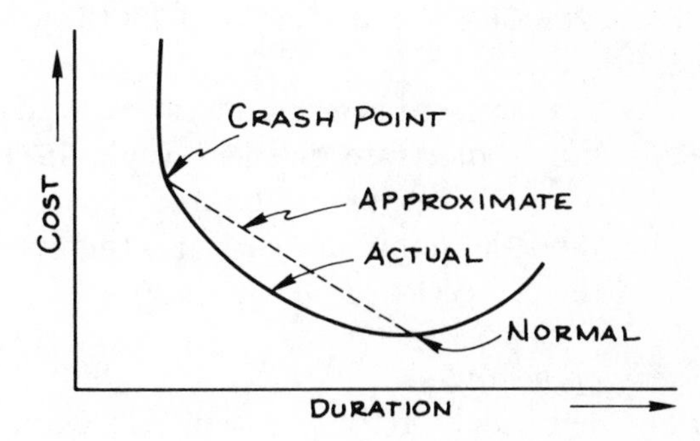

FIGURE 11-8
Duration-cost relationship.

In any construction project, both direct and indirect costs must be considered. Direct costs are those for labor, materials, and plant and equipment; indirect costs include job overhead, taxes, insurance, liquidated damages, and penalties. In many cases it is more economical to crash some activities rather than to pay the indirect costs associated with a long project duration. The additional cost per day or other time unit caused by crashing an activity is the slope of the duration–cost curve, that is, the difference between crash and normal cost divided by the difference between normal and crash times. When this cost for a critical activity is less than the indirect daily cost for the project, it is sensible to compress the time schedule, and the normal time will thus not be the project time having the least cost.

One of the advantages of the arrow diagram over the bar charts used in many construction offices is that it becomes obvious which activities should be crashed. If the time schedule is to be compressed, only those activities on the critical path should be crashed. The main result of crashing activities not on the critical path is a waste of money. All too often when a project is running behind schedule, the contractor will order overtime work on the entire project, in an effort to complete it on time. The waste incurred by this practice is not evident by study of the bar chart, but is more obvious if the critical path is studied. To speed up the completion, only critical activities must be expedited.

EXAMPLE OF COMPRESSING TIME SCHEDULE

The arrow diagram and time-cost data for a small construction project are given in Table 11-3. Each activity requires different resources. Indirect costs will total $200 per day. Find the minimum cost and the associated time required for the project.

Solution:

The slope of the duration-cost curve is obtained by dividing the difference between crash and normal costs by the difference between normal and crash times. The results are shown in Table 11-4.

TABLE 11-3
Duration-Cost Data

Activity	*Normal Duration*	*Crash Duration*	*Normal Cost*	*Crash Cost*
A	4	4	$1,000	$1,000
B	7	4	1,200	2,100
C	5	4	800	875
D	9	7	1,500	1,800
E	5	2	600	900
F	3	2	700	825
G	9	5	400	600
			$6,200	

TABLE 11-4
Cost Slopes

Activity	*Slope*
A	–
B	$300
C	75
D	150
E	100
F	125
G	50

As shown in Figure 11-9, *A-B-D* is the critical path and only those activities should be considered for a crash program. It is impossible to crash *A*, and since it will cost $300 per day to

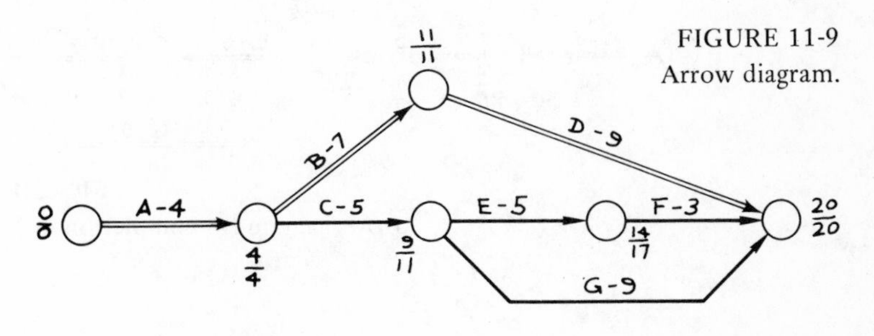

FIGURE 11-9
Arrow diagram.

crash *B*, activity *D* should be considered, since it will cost only \$150 per day extra to perform *D* on a crash basis. If *D* is completed in 7 days, the duration of the project will be 18 days, and the additional cost will be 2 × \$150 = \$300. The arrow diagram will appear then as shown in Figure 11-10.

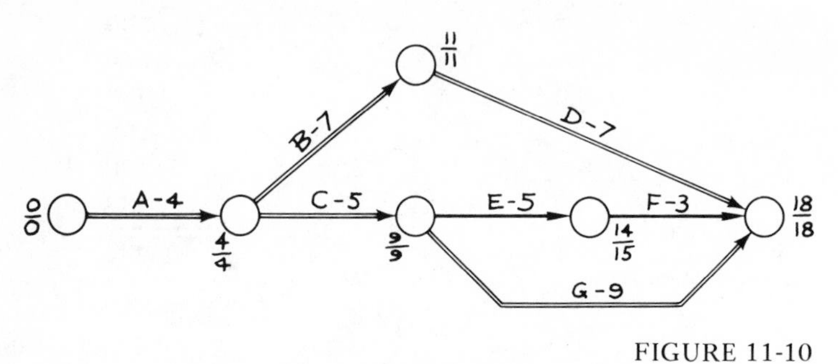

FIGURE 11-10
Arrow diagram for completion in 18 days.

The critical path now has two branches, and in order to reduce the duration of the project, an activity on each branch must be crashed. *D* cannot be reduced further, so by crashing *B*, the cost of the project will be increased by \$300. On the lower branch, crashing *G* will cost \$50, and this will be done rather than crashing *C* at an increase of \$75. The project duration will then be 17 days and the arrow diagram will be as shown in Figure 11-11. All activities are now critical.

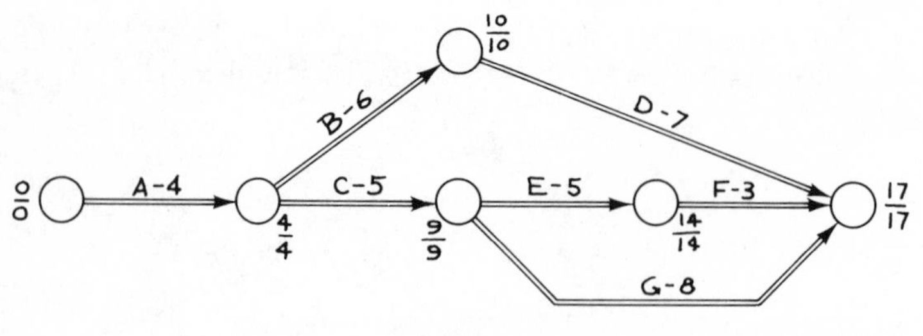

FIGURE 11-11
Arrow diagram for completion in 17 days.

Crashing *B* and *G* again would not save time unless either *E* or *F* were also crashed. The cheapest way to reduce the project duration by one day is to crash *B* and *C*, with an additional cost of \$300 + 75 = \$375. The resulting arrow diagram is shown in Figure 11-12.

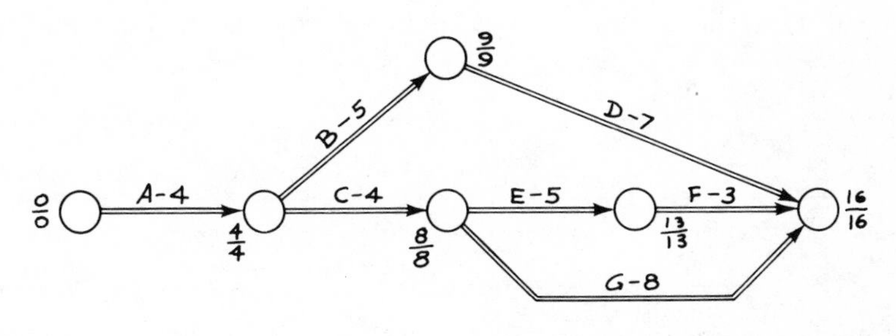

FIGURE 11-12
Arrow diagram for completion in 16 days.

Since *D* has been crashed to its limit, and *B* can be crashed only once more, the minimum completion time is 15 days, accomplished by crashing *B, E,* and *G*. The resulting diagram is shown in Figure 11-13.

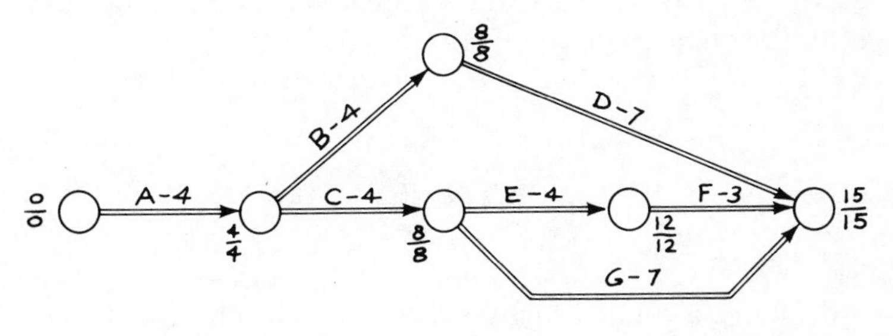

FIGURE 11-13
Arrow diagram for completion in 15 days.

The total project cost is the sum of the direct and the indirect cost and is shown in tabular form in Table 11-5. The total cost decreases as the project duration decreases and reaches its minimum value of $10,100 at 18 days. When the time is reduced below that value, the cost starts to rise again. For minimum cost the project should be scheduled so that its completion time is 18 days, even though additional resources above the normal method of construction are required on some of the activities. A crash program always costs money,

TABLE 11-5

Total Costs with Various Durations

Duration (days)	*Activity Crashed*	*Added Cost*	*Direct Cost*	*Indirect Cost*	*Total Cost*
20	None	–	$6,200	$4,000	$10,200
19	*D*	$150	6,350	3,800	10,150
18	*D*	150	6,500	3,600	10,100
17	*B, G*	350	6,850	3,400	10,250
16	*B, C*	375	7,225	3,200	10,425
15	*B, E, G*	450	7,675	3,000	10,675

but, as seen from this simple example, if crashing some activities costs less than the indirect costs, overtime work or other techniques are justified.

11.4 TIME-SCALED ARROW DIAGRAM

Arrow diagrams are not usually drawn to scale, and the directions of the arrows have no significance. The purpose of the diagram is to show the relationships between the activities, and the placement of the arrows on the diagram should be to make these relationships as clear as possible. Because of its importance, the critical path is often placed at or near the center of the diagram, with the less important activity arrows placed above or below the critical path.

If all the arrows are drawn horizontally, the result is similar to the familiar bar chart used in many construction offices. If arrows are drawn to scale, calendar dates substituted for the number of working days, and dashed lines used to indicate float, the result shown in Figure 11-14 can be very useful for field use. In such a

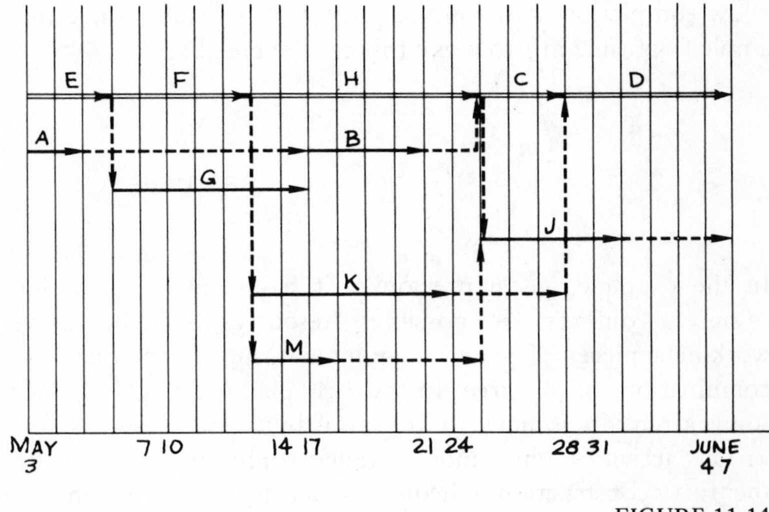

FIGURE 11-14
Time-scaled diagram.

diagram the total float can be added at the end of a series of subsequent activities, or, as in this case, the free float for each activity can be used. Saturdays and Sundays are not usually working days, so they have been omitted. The critical path is shown by doubled lines on the arrows.

A bar chart is very useful in showing at a glance the starting and completion dates for each activity. The arrow diagram shows clearly the relationships between activities but is somewhat difficult to use in determining whether the project is proceeding according to schedule or not. The time-scaled diagram seems to provide the best features of both the bar chart and the ordinary arrow diagram. Referring to Figure 11-14, the project planners decided that activities *A* and *E* could both start on May 3, but if *A* is delayed for some reason, no harm is done if it is completed by the end of May 17. The planners have also decided that K and M can be concurrent, but if they both make use of the same equipment, it should be clear that if M is performed first, followed by K, completion of the project on time will not be difficult, and this change will not affect J. On the other hand, performing K first, then M, will postpone the start of J by 2 days, but will still allow completion of the entire project on schedule, since there is ample float time in J to make this change possible.

11.5 RESOURCE SCHEDULING

In the examples given previously it has been assumed that no resource conflicts are possible. Resources may be crews of workmen, pieces of plant or equipment, money for financing, or combinations of all three. In the early planning stages of a project some attention is paid to the availability of resources for the various activities, but a more detailed study must be made before the final construction schedule is decided upon. When two or more activities are scheduled to be performed concurrently, the demand made upon resources must not be greater than the resources available. If such is the case, one or more activities must be postponed, and unless there is sufficient float for the postponed activities, the duration of the entire project will be increased.

In addition to the obvious resolving of conflicts when two or more activities make use of the same limited resources, it is often desirable to postpone the start of some activities to spread the use of some resources over a longer period of time. The use of a small number of workmen or other resource over a long period is often better practice than using a large number for a short period, then laying off the men or storing the equipment. This leveling of resources can be accomplished by using float time for noncritical activities whenever possible. When time is not critical, it may be desirable to increase the duration of the entire project to avoid peak resource demands.

The bar chart and time-scaled arrow diagram are both very useful in locating resource conflicts. When a conflict is noted, one of the conflicting activities must be postponed. This requires rescheduling all subsequent activities and may cause an increase in the duration of the project. The schedule change that causes the least increase in project duration is usually the most desirable, but determining the best course of action when several activities conflict can be a very tedious task. The use of a computer, with its large memory capacity and ability to do routine calculations very quickly, is ideally suited to resource scheduling, but the process can be performed manually, especially for small projects.

When float time exists for conflicting activities the scheduling problem is not too difficult, but when several activities with little or no float can be performed concurrently and require the same resources, the problem can become a long series of trials. In Figure 11-15 activities x and y are shown on a bar chart. If both activities

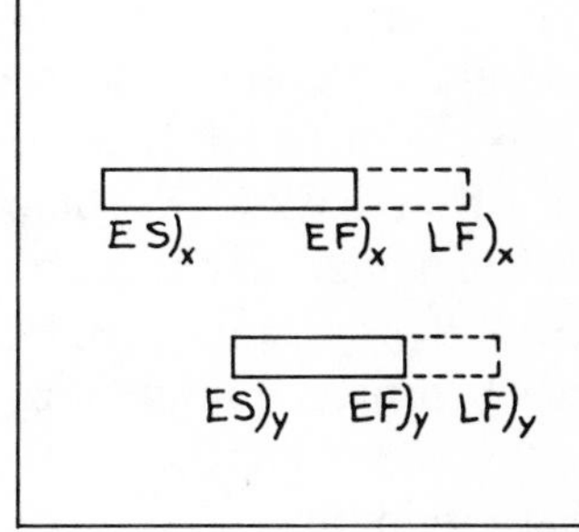

FIGURE 11-15
Activities x and y neglecting resources.

require the same resources, it is desired to determine whether to perform activity x and then y, or complete y first, then do x. $\text{IPD})_{xy}$ is the increase in project duration if x is performed, then y. It is desired to keep IPD to a minimum.

Figure 11-16 shows what will happen if activity x is performed first, then y. $\mathrm{IPD})_{xy}$ is the increase in project duration if x is followed by y.

$$\begin{aligned}\mathrm{IPD})_{xy} &= \mathrm{EF})_x + \mathrm{Dur.})_y - \mathrm{LF})_y \\ &= \mathrm{EF})_x - \mathrm{LF})_y - \mathrm{Dur.})_y \\ &= \mathrm{EF})_x - \mathrm{LS})_y\end{aligned}$$

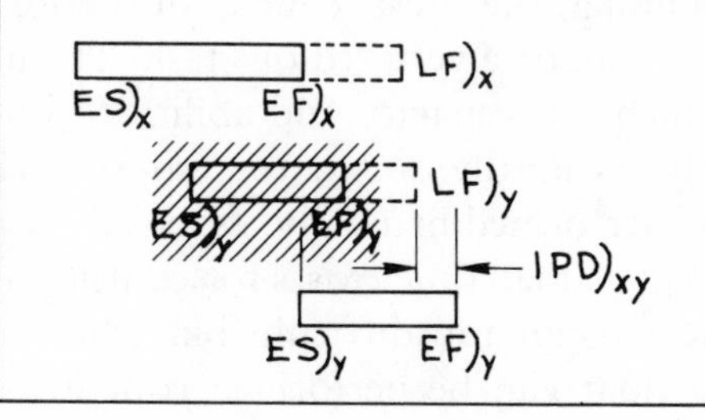

FIGURE 11-16
x and y if y follows x.

When several activities conflict because of the resources demanded, the minimum IPD can be found by using the smallest EF and the largest LS of the conflicting activities to determine the sequence of activities. A computed negative IPD means that because of existing float there will be a zero IPD. The magnitude of negative IPD is meaningless, since the original planning aimed at a minimum project-completion time, and simply causing one activity to follow another cannot reduce the project duration below that minimum.

EXAMPLE OF RESOURCE SCHEDULING

The arrow diagram and tabulated data in Figure 11-17 have been prepared without any consideration of the possibility of resource conflicts. The project manager has decided that the same workmen will be used for both activities E and G; thus they cannot be scheduled at the same time. The same men will be used for *F* and *G* so that these activities cannot be concurrent. Prepare a new schedule for the project.

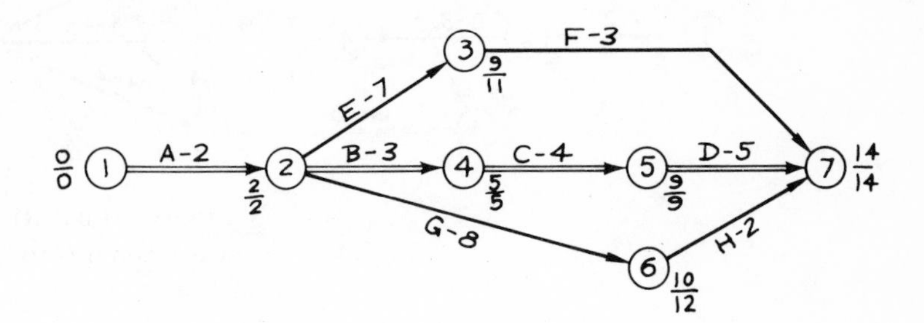

Activity	*Duration*	*ES*	*EF*	*LS*	*LF*	*TF*
A	2	0	2	0	2	0
B	3	2	5	2	5	0
C	4	5	9	5	9	0
D	5	9	14	9	14	0
E	7	2	9	4	11	2
F	3	4	7	11	14	7
G	8	2	10	4	12	2
H	2	8	10	12	14	4

FIGURE 11-17
Arrow diagram and tabulated data.

Solution:
The original plan for the project shows *E* and *G* ready to start after 2 days, but since they will make use of the same men, one must be postponed. If *E* is performed first, $IPD)_{EG} = EF)_E - LS)_G$, or $9 - 4 = 5$ days. If *G* is performed first, $IPD)_{GE} = EF)_G - LS)_E$, or $10 - 4 = 6$ days. Since the smaller IPD is desired, *E* will be performed, then *G*, and the increase

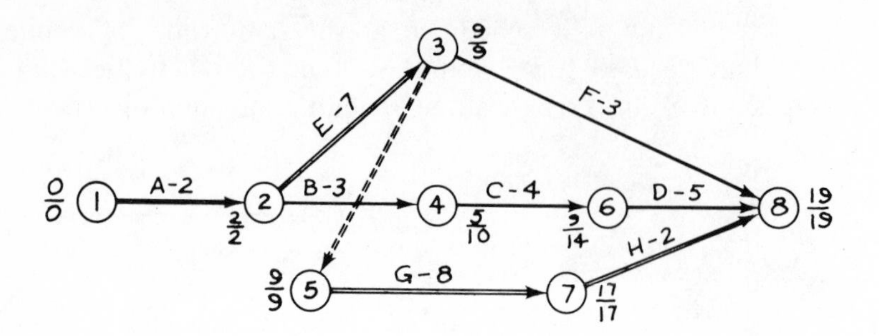

FIGURE 11-18
Arrow diagram if G follows E.

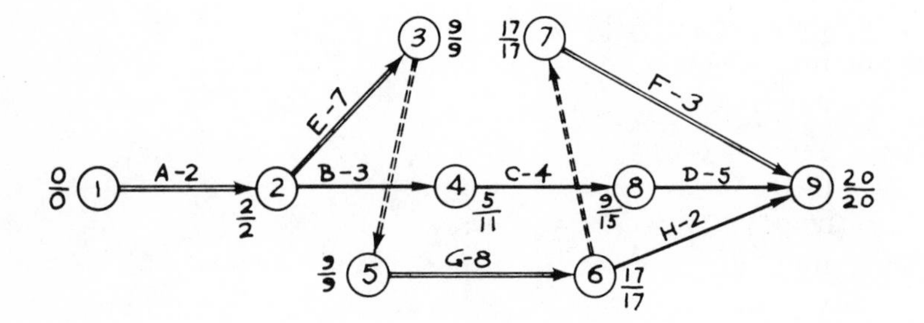

Activity	*Duration*	*ES*	*EF*	*LS*	*LF*	*TF*
A	2	0	2	0	2	0
B	3	2	5	8	11	6
C	4	5	9	11	15	6
D	5	9	14	15	20	6
E	7	2	9	2	9	0
F	3	17	20	17	20	0
G	8	9	17	9	17	0
H	2	17	19	18	20	1

FIGURE 11-19
Scheduling data with conflicts removed.

in project duration will be 5 days. The total duration will then be 14 + 5 = 19 days, and the arrow diagram will be as shown in Figure 11-18. The dummy after *E* has been added to show that *G* must follow *E*, but now a conflict arises between *F* and *G*, since both can start after 9 days. $\text{IPD})_{FG}$ = 12 - 9 = 3 days. $\text{IPD})_{GF}$ = 17 - 16 = 1 day. The total project duration will be 19 + 1 = 20 days, and a dummy must be drawn to show that *F* must follow *G*. The final arrow diagram and tabular data are shown in Figure 11-19.

11.6 UPDATING

Construction projects seldom proceed exactly according to schedule, and a periodic examination of the work might reveal that it is behind schedule. Weather conditions, delays by subcontractors, delays in delivering materials, and revisions in the original planning may require that the schedule be updated. Minor changes and lost time that may be made up are not occasions for updating, and only delays that will cause a significant change in the project completion date should be considered. Updating too frequently may convey the impression that the originally planned deadlines need not be met and no attempt need be made to get the project back on schedule. Updating too infrequently may result in conflicts between subcontractors and other workmen who are trying to adhere to the original schedule.

There are many ways of updating an arrow diagram, but perhaps the simplest is to consider the duration of all completed activities as zero and show the ES for each completed activity as the date of the updating. The duration of each activity in progress at the time should be the actual time required for completion. Durations of all other activities will still be the best possible estimates, whether the same or different from the original estimates of duration. Once the arrow diagram is drawn, calculations of the starting and finishing times can proceed as usual.

11.7 DEFINING ACTIVITIES

An activity or operation is usually thought of as the work performed by a crew of men, or it may be work of a certain nature. What constitutes an activity is somewhat governed by the use that will be made of the planning for a project. If the purpose of the planning is to acquaint top management with what is going on, the presence of great detail should be avoided. Thus, "Building Foundation Walls" as the title of an activity may not be objectionable even though different operations will take place in the building of a wall. The introduction of "Excavating," "Backfilling," "Placing Reinforcing Steel," "Building Forms," "Placing Concrete" as activity titles will add little to the picture desired by management, and indeed will add much confusion.

The arrow diagram used by the project manager in the field should contain as much information as possible in helping him plan his work. If having material on hand for construction of forms is likely to be a problem, ordering the materials should be considered an activity (since it requires time) and treated the same as any other activity. Although of no great interest to management, building forms, placing concrete, placing steel, and so on, are activities important to field management and should be listed as activities and shown on the arrow diagram and other schedules.

Activities that require much time can often be subdivided. In building the foundation walls for a large building, showing that all the excavation must be completed before any of the forms can be started, showing that all forms must be placed before pouring the concrete, and so on, may give an incorrect idea of the field procedure that is to be used and can also indicate a very large duration for building the walls. If carpenters will start building forms for one wall as soon as excavation for that wall is completed, rather than one activity "Excavation," several excavating activities, such as "Excavating West Wall" and "Excavating North Wall" should be used; the activities for forming, placing steel, and the like, should be subdivided in the same manner.

11.8 COMPUTERS

Both CPM and PERT were developed for use with computers, but both can also be used manually. No attempt will be made here to describe the many computer applications, but all computer manufacturers have programs for many features of CPM and PERT.

The arrow diagram must be drawn whether a manual or computer approach is planned, and the analysis of the project resulting in the diagram should be performed whether CPM is to be used or not. This forced, detailed planning has beneficial results even for those contractors who give lip service to CPM and submit arrow diagrams and schedules with their bids because they are forced to, but do not really make use of the method in their later management of the project.

CHAPTER 12

Concrete Forms

The use of reinforced concrete in structures of all sorts is very common and the construction of the forms, or the mold into which the concrete is poured, forms a very important part of the structure. Since the cost of the formwork comprises approximately one-half of the total cost of the work, the economic aspect of this function can be readily appreciated.

The formwork for many structural elements is rather standardized, and very often an experienced carpenter foreman can be expected to design the forms, with little or no supervision, simply by extrapolating previous designs. Many of the details of form building have evolved gradually through the years and the experienced form carpenter will be aware of many features or tricks that the design engineer may not know of.

Concrete forms are typical of many temporary structures that are not given the care in design and construction that they merit. Structural failures are far too common, with a consequent loss of life and property, and too many of these failures are caused by improper fastening and bracing of forms, falsework, scaffolding, and similar temporary structures used in the construction process.

Many carefully made structural designs have resulted in unsatisfactory structures because the construction methods used in the field were left to the guesswork of those not especially well versed in the planning and building of safe and economical structures. Even many experienced field men who have built many successful structures by extrapolating previous designs can be fooled by relatively minor changes in load or dimension.

The requirement that formwork be adequate to carry the imposed loads with safety is obvious, but rigidity of the work may be very important also, but not so obvious. A complete concrete structure should have flat, level floors, and walls should be straight or uniformly curved as planned by the designer. The dimensions of all structural elements should be as planned, and in a concrete structure all sizes and shapes of the finished product are controlled by the forms. The forms must therefore be rigid enough to withstand all impact loads and vibrations during construction without any important change in position or dimension. Adequate connections of form elements as well as sufficient bracing are sometimes overlooked or greatly underdesigned for rigidity as well as strength and account for many collapses during or after concreting operations.

The designer of formwork must be extremely cost conscious if concrete is to compete successfully with other building materials. The quantity of concrete and reinforcing steel in a structure will be governed by the designer, but the material and labor costs involved in making, erecting, and removing the forms are the province of the form designer, and substantial savings can be made by the proper choice of methods and materials.

The material chosen for the forms should be such that it may be used and reused as many times as possible, to keep the cost per use at a minimum. The oiling or wetting procedure used before concreting will affect the ease of stripping and consequent damage to the form material. The method of cleaning and storage after each use will also have an effect upon the damage to the material and the possibility of further use. Any cleaning, repairing, or other treatment of the concrete after stripping the forms is costly, and can be kept at a minimum by the proper choice of form materials. When not required for strength or smoothness of finish, the use of low-grade and inexpensive lumber can often result in substantial savings.

As much carpentry as possible should be performed with power machinery and under the best possible working conditions. This usually means that relatively large units or panels are prefabricated in the carpenter shop and erected at the site. Considerable judgment must be exercised in the size of the panels since their weight must be balanced by the capacity of the hoisting equipment available. If large-capacity cranes are available at the site, the use of relatively large panels is indicated for efficiency of the cranes. The procedure to be used in stripping the forms after the concrete sets should also be considered in determining the most economical size of the panels. Panels of a given size may be ideally suited to the available hoisting equipment for erecting, but difficulties in removing them might tie up the use of a crane long enough to be uneconomical.

Wasting of lumber can be prevented by the use of standard-sized panels for many wall or floor sections with a minimum of alterations. Minimum curing time will permit early stripping of forms and their subsequent use elsewhere. The use of high-early-strength cement combined with optimum curing conditions of temperature and humidity, although costly in themselves, will sometimes more than pay for themselves in savings of form materials.

12.1 FORM MATERIALS

For many years most forms were made from inexpensive grades of softwood boards. For rough work, butt-jointed boards are satisfactory, while tongue-and-groove boards are often used when a better surface is required, or when vibrations or erection stresses might cause openings between adjacent boards to occur. Large quantities of lumber are still used for forms, but the use of boards or tongue-and-groove lumber for large surfaces is decreasing.

Plywood has largely replaced ordinary lumber for forming large surfaces—either flat or curved. It can be fabricated into panels of any desired size and can be used many times if care is taken during erecting and stripping. The forms may be lined with plastic or metal to achieve different finishes in the concrete, or many decorative or unusual effects can be made by using different grades or species of plywood.

Metal forms, usually of steel, but sometimes of aluminum, are often used when very large and regular surfaces are to be formed. They are expensive to purchase or rent, but since they can be used many times without damage, the cost per use may be quite low. Such forms are often used for walls or floors and the panels are bolted together or fastened with patented clamps of many sorts.

12.2 DESIGN CONSIDERATIONS

Formwork must be designed to support the concrete and any other loads imposed on the structure without undue deformation until the concrete has set and can carry the load itself. Whether the structural element be a column, wall, or floor slab, the loads can be considered as either dead or live loads.

Dead loads are usually considered to be the weight of the structure itself. In most structures dead loads can be calculated rather precisely. In the case of forms, once the size of the members has been determined and the species of lumber or other material decided upon, the weight can be found, but since form weight is so small compared with the loads supported, this dead load is sometimes neglected.

Live loads consist of the weight of concrete and any forces associated with placing it, reinforcing steel and accessories, and the weight of any equipment and workmen used in placing the concrete, including any possible impact. Values of live loading on a slab to represent workmen, equipment, and impact should be at least 40 lb/ft^2. A value of 50 lb/ft^2 is common if nonmotorized equipment is used, and 75 lb/ft^2 if motorized buggies are used in placing the concrete. If extensive use of the green concrete for storage of equipment or materials is to be made, the actual weights anticipated must be applied.

Different concrete mixes and steel quantities will cause the unit weight of reinforced concrete to vary somewhat, but a value of 150 lb/ft^3 is usually used for ordinary concrete, including reinforcing. The use of slag, cinders, and so on, as the aggregate will make lighter concrete such that smaller weights, such as 100 or 120 lb/ft^3, are used, depending upon the particular aggregate used.

Wall and column forms must be designed to withstand a wind load of at least 10 lb/ft^2 unless a higher value is required because of the location or requirements of a local building code. Forms over 8 ft tall are also sometimes required to be designed to withstand 100 lb/linear ft applied at the top of the form.

When concrete is placed in wall forms it exerts a force perpendicular to the surface of the forms. The magnitude of this lateral pressure has been and continues to be the subject of much research, with varying results. Fresh concrete is a fluid, and the lateral pressure it exerts is proportional to its depth. As the concrete sets, the pressure due to the initial hydrostatic head is reduced and the concrete, when partially set, exerts very little lateral pressure on the forms. The horizontal force exerted on a wall or column form thus varies with the condition of the concrete, from nearly zero to the full hydrostatic pressure. Some of the factors pertaining to the condition of the concrete must therefore be examined.

The unit weight of concrete, the water–cement ratio, and the proportions of fine and coarse aggregate will have some effect upon the lateral pressure developed, but a mix of average workability with a slump of about 4 inches and the generally used weight of 150 lb/ft^3 will be assumed. The use of vibrators is so universal that it will be assumed that internal vibrators are always employed.

If concrete is poured into wall forms so slowly that the concrete in the bottom layer has partially set before the next layer is poured, there will be no appreciable hydrostatic pressure at any depth. When the top layer of concrete is poured, the hydrostatic head would be due to the thickness of that layer only, and because they were partially set, the bottom layers would be self-supporting and would require little help from the forms. Since the forms are needed only while the concrete is acquiring some of its initial set, any conditions relating to the amount of set are important. The two most important conditions are the rate of fill and the temperature. A high rate of fill will not allow sufficient set to take place in the lower layers before the top layers are poured, and a high pressure would thus be present at all depths and must be designed for. The reverse, of course, would be true for a low rate of fill.

For rapid curing of concrete, high temperature and high humidity are required. Humidity control in the early stages of curing is easy to attain and will be assumed. Optimum curing temperature is approximately 90°F, and curing time will be increased when the temperature is either higher or lower than that.

The relationships between the factors involved are complex and many experimenters feel that they are overly conservative, but the American Concrete Institute recommends the following values.

For columns with internal vibration,

$$P = 150 + 9{,}000\,\frac{R}{T} \quad \text{(with a maximum of 3,000 lb/ft}^2\text{ or }150h\text{, whichever is least)} \qquad \text{(Eq. 12-1)}$$

For walls with internal vibration,

$$P = 150 + 9{,}000\,\frac{R}{T} \quad \text{(with a maximum of 2,000 lb/ft}^2\text{ or }150h\text{, whichever is least)} \qquad \text{(Eq. 12-2)}$$

For walls with R greater than 7 ft/hr,

$$P = 150 + \frac{43{,}400}{T} + 2{,}800\,\frac{R}{T} \quad \text{(with a maximum of 2,000 or }150h\text{, whichever is least)} \qquad \text{(Eq. 12-3)}$$

where P = lateral pressure, lb/ft^2
R = rate of fill, ft/hr
T = temperature, °F
h = height of fresh concrete above the point considered, ft

Building codes and standard specifications contain working stresses for the various materials used in construction. Lumber and plywood stresses as usually given are for dry conditions and loads of long duration. Wood, when wet, is weaker than dry wood, so the usual working stresses must be reduced somewhat for wood used in forms, since forms are lubricated with oil or kept wet with water for ease in stripping, and of course they are in contact with wet concrete. The loads imposed on forms are for a short duration, and for this situation an increase in stress is permitted. The result of these two adjustments permits working stresses for formwork somewhat higher than those for dry conditions and permanent loading.

Any structure will deflect when loaded, and of course formwork is no exception. The characteristics of the structure will govern the deflection permitted, with deflections of 1/8 inch or 1/360 times the span, whichever is smaller, being typical of the magnitude of sag allowed by many specifications.

12.3 TYPICAL FORMS

Entire books have been written on the design of forms for concrete structures and have done little more than to introduce the student to some of the theory and art of design, leaving many of the tricks and practical aspects to be learned from experience. This one chapter can do no more than to mention the more obvious and simple forms, leaving the complicated systems and the many excellent patented devices for the student who wishes to specialize in this phase of construction.

Forms for large footings and ordinary walls are similar and may be made of steel, ordinary lumber, plywood, or a combination of materials. A common and typical one, shown in Figure 12-1, consists of plywood or lumber sheathing, studs spaced

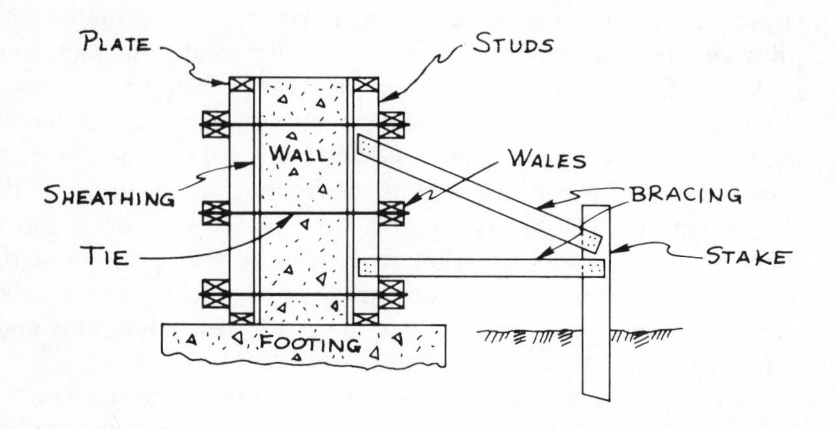

FIGURE 12-1
Typical wall forms.

uniformly between 1 and 3 feet apart along the length of the wall, double wales or walers parallel with the long dimension of the wall, and bracing at intervals of 6 to 8 feet along the length of the wall. The walls are held together by metal ties or wires which are cut or twisted off back from the concrete surface, and the resulting holes in the concrete plugged or otherwise patched. Wales are usually double to facilitate installation of the wedges or other connections at the ends of the ties.

A basic building block for forming walls is a 4- × 8-foot sheet of plywood reinforced with 2 × 3's or 2 × 4's as shown in Figure 12-2. Panels can be nailed, bolted, or clamped together as desired, and made rigid by wales and form ties.

Round columns are usually formed by heavy wood fiber or cardboard tubes which are available in many diameters and lengths. These tubes are manufactured by gluing together large sheets of fiber, either spirally wound or seamless. When stripped, the concrete is smooth and requires little or no surface treatment.

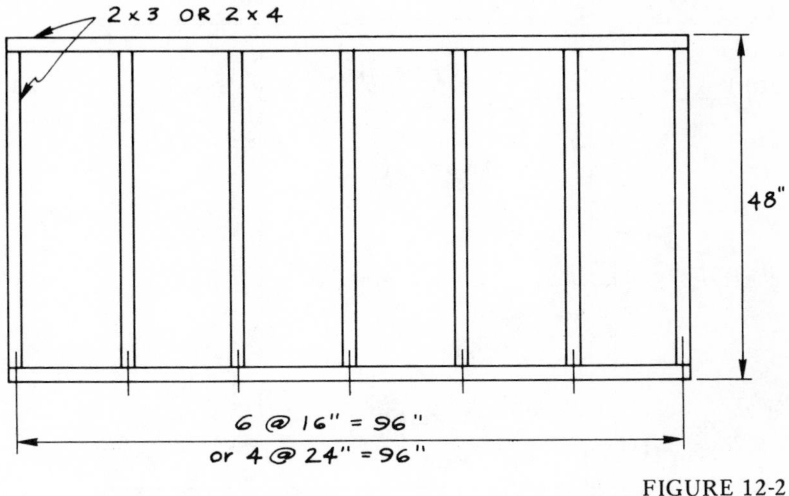

FIGURE 12-2
Plywood form panel.

Columns of rectangular cross section are sheathed with plywood or plain lumber. The desired dimensions of the column are assured by the use of wooden yokes or adjustable steel clamps spaced 1 foot or more apart vertically. Many types of steel clamps may be rented or purchased and have largely supplanted the use of wooden yokes.

The forms for a common beam and slab floor system are shown in Figure 12-3. For the slab, the board or plywood deck is supported by 2-inch joists spaced 1–3 feet apart, and these in turn are supported by larger members called stringers, which are spaced several feet apart. If the span is large enough, the stringers are supported by intermediate shores or posts. For short spans, the stringers frame into and are supported by the forms for the beams, T-heads have been used for many years to support the forming for the beams, but much tubular steel scaffolding which can be rented or purchased is now used for the vertical support of both beams and slabs.

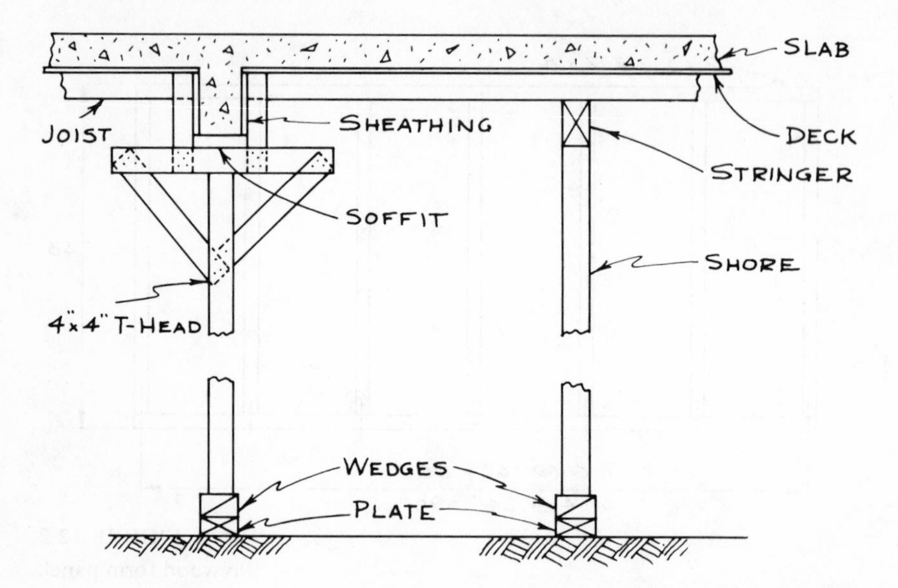

FIGURE 12-3
Forms for beam and floor slab.

12.4 FORM DESIGN

Most formwork is statically indeterminate to a large degree, but by making simplifying assumptions, much design can be performed by using the ideas and techniques learned in a beginning course in strength of materials. Most components of a form system act as beams, and in the design of a beam the bending stresses, shearing stresses, and deflection must be investigated. Vertical shores and other support members act as columns and are designed as such.

In a beam, the allowable bending stress is

$$F_b = \frac{Mc}{I} \qquad \text{(Eq. 12-4)}$$

where

M = bending moment
c = distance from the neutral axis to the extreme fiber measured on the cross section
I = moment of inertia of the cross section

If the cross section is rectangular, b inches wide and h inches high,

$$c = \frac{h}{2} \quad \text{and} \quad I = \frac{1}{12}bh^3$$

Substituting these values into the flexure formula as given in Equation 12-4,

$$F_b = \frac{6M}{bh^2} \qquad \text{(Eq. 12-5)}$$

In shear, the allowable stress is

$$F_v = \frac{VQ}{Ib} \qquad \text{(Eq. 12-6)}$$

where V = external shearing force
Q = moment of area of the cross section beyond the point of maximum shear with respect to the neutral axis
I = moment of inertia of the cross section
b = beam width at the point of maximum shear

For a rectangular cross section,

$$Q = \frac{bh}{2}\frac{h}{4} = \frac{bh^2}{8} \quad \text{and} \quad I = 1/12bh^3$$

Substitution of these values into the general shear formula gives

$$F_v = \frac{3V}{2bh} \qquad \text{(Eq. 12-7)}$$

The deflection of a beam acted upon by a uniformly distributed load of w lb/ft is

$$\Delta = \frac{wL^4}{K_1 EI} \qquad \text{(Eq. 12-8)}$$

where L = span
E = modulus of elasticity
K_1 = constant and depends upon end conditions of the beam or

$$\Delta = \frac{PL^3}{K_2 EI} \qquad \text{(Eq. 12-9)}$$

where P = concentrated load or force
K_2 = constant and depends upon end conditions

Columns can be considered as either long or short, depending upon their length with respect to the smaller cross-sectional dimension, and design formulas are different for each category. If

$$L/d \leqq \sqrt{\frac{0.3E}{F_c}} \qquad \text{(Eq. 12-10)}$$

where F_c = allowable stress in compression parallel to the grain
L = unsupported length of the column
d = least cross-sectional dimension

the result is defined as a short column. If the reverse is true, the long-column formula applies. For a short column, buckling is not critical and the allowable load is

$$P = F_C A \qquad \text{(Eq. 12-11)}$$

where A is the cross-sectional area. For a long column,

$$P = \frac{0.3EA}{(L/d)^2} \qquad \text{(Eq. 12-12)}$$

EXAMPLE OF FLOOR-SLAB FORM DESIGN

Design the forms for a large floor slab. Slab thickness is to be 6 in. and a live load of 50 lb/ft^2 must be provided for. A maximum deflection of 1/8 in. for each form component is specified.

Solution:
The working stresses to be used are for no particular species but are typical of many structural grades and are as follows:

F_b = 1,600 lb/in.^{2} for bending

F_v = 150 lb/in.^{2} for shear

$F_{c\perp}$ = 400 lb/in.^{2} for compression perpendicular to the grain

F_c = 1,500 lb/in.^{2} for compression parallel with the grain

E = 1,600,000 lb/in.^{2} as a modulus of elasticity

There are several methods that might be used. In this example lumber sizes will be chosen and the spacing of supports calculated. The resulting design may not be the best possible. The optimum design must be determined by comparing the results of several designs and choosing the most economical. As a practical matter, the lumber sizes available must be considered since often the most economical design may call for sizes of material that are not on hand.

For the decking, nominal 1-in. (3/4-in. actual) tongue-and-groove boards of variable length will be used. Most of the deck boards will be continuous over three or more spans; therefore, the maximum bending moment, M, will be $1/10wL^2$ lb-ft or $1/10wL^2 \times 12$ lb-in. if the load, w, is in lb/ft and the length of span, L, is in ft. Substituting this value for M into Equation 12-5 gives

$$F_b = \frac{6}{bh^2} \quad \frac{wL^2 \times 12}{10} \qquad \text{(Eq. 12-13)}$$

Solving,

$$L = \sqrt{\frac{F_b bh^2}{7.2w}}$$

The decking can be thought of as a beam 3/4 in. high and 12 in. wide. Its length is unknown and is the distance between supporting joists. The load w, in lb/ft of beam, is the weight of the concrete supported plus the live load, hence

$$w = \frac{6\text{in. thick} \times 12 \text{ in. wide} \times 1 \text{ ft} \times 150 \text{ lb/ft}^3 + 50 \text{ lb/ft}^2}{144 \text{ in.}^2/\text{ft}^2}$$

$$= 75 + 50 = 125 \text{ lb/ft}$$

With $F_b = 1{,}600$ lb/in.^{2} from Equation 12-13,

$$L = \sqrt{\frac{1{,}600 \times 12 \times (3/4)^2}{7.2 \times 125}} = 3.47$$

$$= \text{joist spacing based on bending}$$

For shear, the maximum of V, the shearing force, will occur in a two-span beam and equals $5/8wL$. Substituting this value in Equation 12-7,

$$F_v = \frac{3}{2bh} \quad \frac{5wL}{8} = \frac{15\,wL}{16\,bh} \qquad \text{(Eq. 12-14)}$$

Solving,

$$L = \frac{F_v \cdot 16bh}{15w}$$

Substituting numerical values,

$$L = \frac{150 \times 16 \times 12 \times 3/4}{15 \times 125} = 11.5 \text{ ft}$$

$$= \text{joist spacing based on shear}$$

The deflection of a beam as given in Equation 12-8 is very general and the constant K_1 must be determined for each case in which it is used. The decking of a slab will be continuous; that is, most boards will extend over several supports. It is possible, however, especially near the ends of the slab, that short lengths of board will be used. Any discontinuity in the slab for stairways, crane or hoist openings, and so on, might also allow the use of short boards. A very conservative design would treat the decking as a simply supported beam between adjacent joists, and this approach will be used here. The value of K_1 will then be 5/384 and Equation 12-8 becomes

$$\Delta = \frac{5wL^4}{384EI} \times 1{,}728$$

The conversion factor of 1,728 in.^{4}/ft^4 is necessary if the deflection is expressed in inches. Solving,

$$L^4 = \frac{5{,}920\Delta bh^3}{w} \qquad \text{(Eq. 12-15)}$$

or

$$L = \left[\frac{5{,}920 \times 1/8 \times 12 \times (3/4)^3}{125}\right]^{1/4} = 2.34 \text{ ft}$$

$$= \text{joist spacing based on deflection}$$

The design of the decking is governed by deflection rather than strength, since spans of 3.47 and 11.5 ft respectively, for bending and shear are larger than the 2.34 ft required by the 1/8-inch specification for deflection. Any convenient distance less than 2.34 ft can be used for the decking span (joist spacing), and in this case 2 ft will be specified.

The decking will be supported by joists, and the design of these members will involve the same problems as the decking. If it is decided, somewhat arbitrarily, that 2 × 6's (actually 1.5 × 5.5 in.) will be used, the length of each joist span or the distance between stringers must be determined. As shown

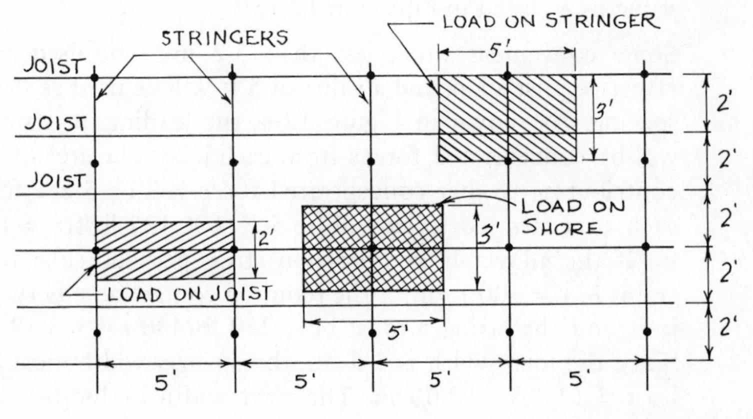

FIGURE 12-4
Tributary loading areas for floor slab.

in Figure 12-4, each linear foot of joist supports a slab 2 ft wide; thus the loading on the joist will be 2 × 125 = 250 lb/ft. Using the same procedure as before, from Equation 12-13,

$$L = \sqrt{\frac{F_b bh^2}{7.2w}} = \sqrt{\frac{1{,}600 \times 1.5 \times 5.5^2}{7.2 \times 250}} = 6.35 \text{ ft for bending}$$

From Equation 12-14,

$$L = \frac{F_v \cdot 16bh}{15w} = \frac{150 \times 16 \times 1.5 \times 5.5}{15 \times 250} = 5.29 \text{ ft for shear}$$

From Equation 12-15,

$$L = \left(\frac{5920 \Delta bh^3}{w}\right)^{1/4} = \left(\frac{5{,}920 \times 1/8 \times 1.5 \times 5.5^3}{250}\right)^{1/4}$$

$$= 5.20 \text{ ft for deflection}$$

It is unlikely that short pieces of lumber will be used for the joists, so the possibility of single-span loading is quite remote. The calculation for deflection span is thus very conservative, but since the numerical values for span based on shear and deflection are so close, little will be gained by adjusting the value of K_1 used in Equation 12-15.

Some convenient value less than 5.2 must be used as the effective joist span, and a value of 5 ft will be used as stringer spacing. As shown in Figure 12-4, the loading on a stringer will be concentrated forces from each joist. The area of floor contributing to that concentrated force will be 5 × 2 ft, and each concentrated force will be 5 × 2 × 125 lb/ft^2 = 1,250 lb. If the allowable compression stress perpendicular to the grain, $F_{c\perp}$ = 400 lb/in.^{2}, the required contact area between a joist and the stringer must be 1,250 lb/400 psi = 3.12 in.^{2}. Since the joist width is 1.5 in., the stringer width must be at least 3.12/1.5 = 2.08 in. The next width of lumber larger than 2 in. that is commonly avialable is 4 in.; therefore, a nominal 4 × 4 (3.5 × 3.5) will be used for the stringers.

Although the load on the stringers is actually a series of concentrated forces from the joists, no appreciable error will result if an equivalent uniform load is used. The area contributing to the stringer load is 5 ft wide; therefore, a load of 5 × 125 = 625 lb/ft² will be used. The length of stringer between supports will be as determined as before, considering bending, shear, and deflection. From Equation 12-13,

$$L = \sqrt{\frac{F_b bh^2}{7.2w}} = \sqrt{\frac{1{,}600 \times 3.5 \times 3.5^2}{7.2 \times 625}} = 3.91 \text{ ft for bending}$$

From Equation 12-14,

$$L = \frac{F_v \cdot 16bh}{15w} = \frac{150 \times 16 \times 3.5 \times 3.5}{15 \times 625} = 3.14 \text{ ft for shear}$$

From Equation 12-15,

$$L = \left(\frac{5{,}920\Delta bh^3}{w}\right)^{1/4} = \left(\frac{5{,}920 \times 1/8 \times 3.5 \times 3.5^3}{625}\right)^{1/4}$$

$$= 3.66 \text{ ft for deflection}$$

In this case shear is critical, so shores should be spaced at 3 ft center to center to act as support for the stringers. The load on each shore or supporting post will come from a floor area 5 × 3 ft, or the vertical force per shore will be 5 × 3 × 125 = 1,875 lb. If a 4 × 4 shore (3.5 × 3.5 inches actual size) is used, the maximum force tending to crush the stringer at the point of support will equal the area of contact multiplied by the allowable stress perpendicular to the grain, or $P = AF_{c\perp}$ = 3.5 × 3.5 × 400 = 4,900 lb. Since 4,900 is much greater than the actual 1,875 lb load, crushing of the stringer is not critical.

If the height or unsupported length of the shore is 10 ft, the slenderness ratio,

$$L/d = \frac{10 \times 12}{3.5} = 34.3$$

From Equation 12-10,

$$\sqrt{\frac{0.3E}{F_c}} = \sqrt{\frac{0.3 \times 1{,}600{,}000}{1{,}500}} = 17.8$$

and since 34.3 is greater than 17.8, a 10-ft shore would be classified as a long column, and, from Equation 12-12,

$$P = \frac{0.3EA}{(L/d)^2} = \frac{0.3 \times 1{,}600{,}000 \times 3.5^2}{34.3^2} = 5{,}000 \text{ lb}$$

Since 5,000 lb is the maximum allowable load on the shore while a load of only 1,875 lb is anticipated, a nominal 4 × 4 shore is satisfactory. The entire design is summarized in Figure 12-5.

There is no unique, correct answer to this simple design problem, and this solution may be unsatisfactory in some respects. Thicker decking, of course, would call for fewer joists, and larger stringers could reduce the large number of shores required. The use of patented tubular steel scaffolding

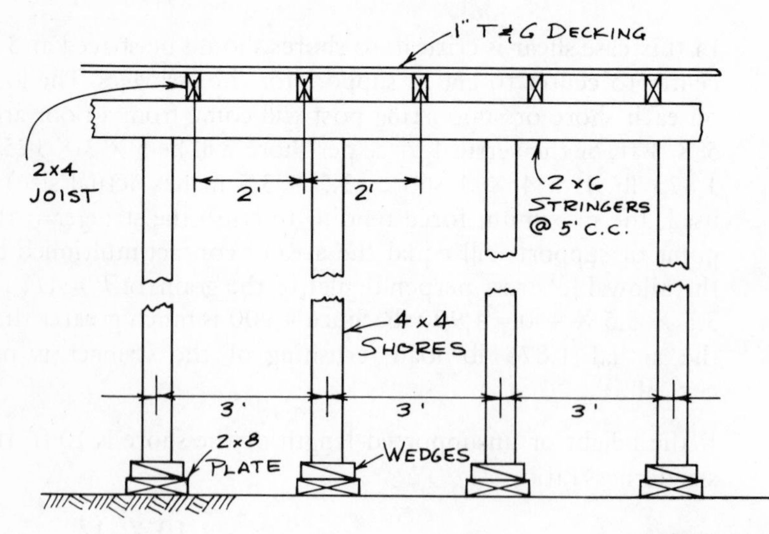

FIGURE 12-5
Forms for floor slab.

would result in a different spacing of stringers, and the availability of different sizes of lumber might dictate a completely different design. The shores should be braced in both directions with either horizontal boards or 1-inch boards applied diagonally, but in alternate bays or every third bay. If any traffic must use the space under the slab before the forms are stripped, the close spacing of shores would be intolerable, and a different design must be used.

EXAMPLE OF WALL FORM DESIGN

Design the forms for a concrete wall 12 ft high and 8 in. thick. Concrete will be placed at the rate of 3 ft/hr and at a temperature of 60°F. The forms will consist of plywood sheathing, 2 X 6 studs spaced 16 in. center to center, and double wales of a size to be determined. The construction will be as shown in Figure 12-1. For lumber, the working stresses used in Example 1 will be used. For plywood the recommendations of the American Plywood Association will be used.

Solution:
The maximum lateral pressure from Equation 12-2 is

$$P = 150 + 9{,}000\,\frac{R}{T} = 150 + 9{,}000 \times \frac{3}{60} = 150 + 450 = 600\ \text{lb/ft}^2$$

or $F = 150h = 150 \times 12 = 1{,}800$ lb/ft^2. The smaller of these values, 600 lb/ft^2, will be used. For sheathing, the recommendations of the American Plywood Association will be followed for beams of three spans. The maximum moment, $M = 1/10wL^2$ if w is in lb/ft and L is in feet but $M = wL^2/120$ if L is in inches. Substituting this value in the flexure formula, Equation 12-4,

$$F_b = \frac{Mc}{I} = \frac{M}{KS}$$ where KS is a modified section modulus

$$F_b = \frac{wL^2}{120KS}$$

or

$$KS = \frac{wL^2}{120F_b} \qquad \text{(Eq. 12-16)}$$

For shear, the maximum $V = 6/10\ wL = 6/120\ wL$ if L is in inches. Substituting into Equation 12-6,

$$F_v = \frac{VQ}{Ib} = \frac{wL}{20}\frac{Q}{Ib}$$

Solving,

$$\frac{IQ}{b} = \frac{wL}{20F_v} \qquad \text{(Eq. 12-17)}$$

or

$$\frac{IQ}{b} = \frac{wL}{20F_s}$$

where F_S is the more applicable allowable stress in rolling shear. Similarly, for deflection,

$$\triangle = \frac{0.0069wL^4}{EI} \quad \text{or} \quad \frac{0.0069wL^4}{12\ EI} \quad \text{if } L \text{ is in inches}$$

or,

$$\triangle = \frac{wL^4}{1{,}743EI} \qquad \text{(Eq. 12-18)}$$

The modulus of elasticity of Class I Plyform is 1,760,000 and working stresses in bending and rolling shear, F_b and F_S are, respectively, 2,010 and 76 psi.

Properties of constants associated with the cross section of plywood will be different from those made from solid lumber because of the orientation of the grain in the various layers of wood making up the plywood. For various thicknesses these effective constants are as shown in Table 12-1.

TABLE 12-1

Thickness (in.)	*I*	*KS*	*Ib/Q*
1/2	0.080	0.273	5.215
5/8	0.134	0.363	6.654
3/4	0.203	0.461	8.192
7/8	0.304	0.591	8.420
1	0.436	0.741	9.136

In bending, from Equation 12-16,

$$KS = \frac{wL^2}{120F_b} = \frac{600 \times 16^2}{120 \times 2{,}010} = 0.636 \text{ in.}^3 \text{ required}$$

In shear, from Equation 12-17,

$$\frac{Ib}{Q} = \frac{wL}{20F_s} = \frac{600 \times 16}{20 \times 76} = 6.33 \text{ in.}^2 \text{ required}$$

For deflection, the value determined from Equation 12-18 can be used after the size of sheathing has been determined. If the wall is a foundation wall, not visible in the completed structure, the value of the deflection may not be critical and might not be computed by many designers. In this case, however, the magnitude of deflection will be specified as no more than 1/360 of the span. Solving Equation 12-18 for I,

$$I = \frac{wL^4}{\triangle 1{,}743E} \qquad \text{(Eq. 12-19)}$$

or

$$I = \frac{600 \times 16^4}{16/360 \times 1{,}743 \times 1{,}760{,}000} = 0.288 \text{ in.}^4 \text{ required}$$

The sheathing chosen must be thick enough that the requirements for bending, shear, and deflection are met. For bending, a KS value of 0.741 furnished by 1-in. plywood is greater than the value of 0.636 in.^{3} required. The Ib/Q of

6.654 furnished by a 5/8-in. thickness exceeds the 6.33 in.^{2} required, and the I of 0.304 furnished by 7/8-in. plywood exceeds the 0.288 in.^{4} required. Bending is therefore critical and the 1-in. thick plywood will be used. With 2 $\times$ 6 studs spaced 16 inches center to center, the loading will be 16/12 $\times$ 600 = 800 lb/ft^2. From Equation 12-13, for bending

$$L = \sqrt{\frac{F_b bh^2}{7.2w}} = \sqrt{\frac{1{,}600 \times 1.5 \times 5.5^2}{7.2 \times 800}} = 3.55 \text{ ft} = \text{stud span}$$

Since the studs will undoubtedly be the full height of the wall in length, the maximum shearing force for three spans is $6/10wL$. Using this value rather than $5/8wL$ in deriving Equation 12-14, this equation becomes

$$L = \frac{20bhF_v}{18w}$$

or

$$L = \frac{20 \times 1.5 \times 5.5 \times 150}{18 \times 800} = 1.72 \text{ ft} = \text{stud span for shear}$$

Of these two values, 1.72 ft or 20.8 in., is critical and represents the clear span of the studs. Wales are usually doubled and the space between blocked such that the total width is at least 4 in. The center-to-center spacing of the wales can be 24 in. and still have a clear stud span of 20.8 in., and this spacing will be used. Deflection will probably not be critical, but from Equation 12-18,

$$\Delta = \frac{wL^4}{1{,}743EI} = \frac{800 \times 24^4}{1{,}743 \times 1{,}760{,}000 \times 1/12 \times 1.5 \times 5.5^3}$$

$$= 0.0041 \text{ in.}$$

This deflection is very small compared with 1/360 $\times$ span or 24/360 = 0.067 in.

Form ties are usually proprietary products supplied with special clamps or wedges so that the required wall thickness

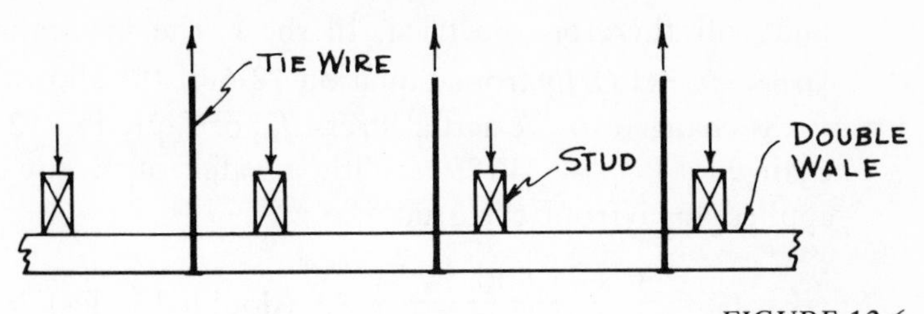

FIGURE 12-6
Loading on wales for wall form.

can be maintained. Most types are used with double wales and provide support horizontally for these members. Tie strengths vary from about 3,000 lb to several times that amount. Tables supplied by manufacturers give the required spacing for different sizes, and no attempt will be made in this example to select a size or spacing. The wales will act as long and continuous beams loaded as shown in Figure 12-6. With studs 16 in. center to center and wales 24 in. apart, a wall area of 16/12 X 2 will contribute to the load at each contact point between a stud and a wale, or these concentrated forces will be 1.33 X 2 X 600 = 1,600 lb. If the tie wires did not provide sufficient support for full continuity of the wales, the worst possible loading condition for the wale would occur when a stud fell midway between two tie wires, and a simple beam with a concentrated force at midspan would be approximated. In this case,

$$M = \frac{PL}{4} = \frac{1{,}600 \times 2 \times 12}{4} = 9{,}600 \text{ lb-in.}$$

From the flexure formula, Equation 12-4, $F_b = Mc/I = M/S$, where $S = I/c$, the section modulus. Solving for S,

$$S = \frac{M}{F_b} = \frac{9{,}600}{1{,}600} = 6 \text{ in.}^3\text{, the required section modulus}$$

Two 2 X 6's would furnish a value of

$$S = \frac{I}{c} = 2\ \frac{1/12 \times 1.5 \times 5.5^3}{5.5/2} = 15.12$$

and will therefore be tried. In shear, the maximum actual stress, $fv = VQ/Ib$ from Equation 12-6, if the allowable shear F_v, is changed to the actual stress, f_v, or Equation 12-7 can be changed to $f_v = 3V/2bh$. With a value of V equal to the applied load from the stud,

$$f_v = \frac{3 \times 1{,}600}{2 \times 1.5 \times 5.5} \times 2 \text{ (doubled wales)} = 145 \text{ psi.}$$

Since the allowable stress, F_v, = 150 psi, double 2 × 6 wales are satisfactory in both bending and shear and will be used. Deflection is seldom a problem in either studs or wales and will therefore not be checked numerically.

The final wall design is as shown in Figure 12-7.

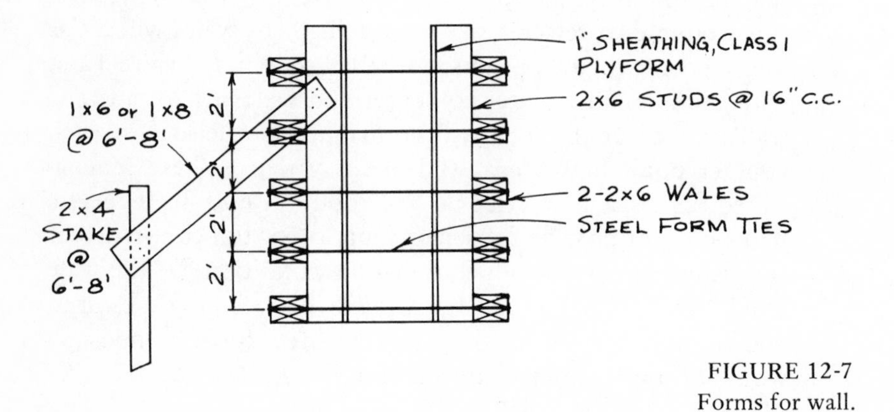

FIGURE 12-7
Forms for wall.

12.5 FAILURE OF FORMWORK

Construction literature all too frequently reports on the failure of concrete forms during construction. The causes of these failures can sometimes be determined, but often the failure of one of the components sets up a chain reaction and in the resulting rubble it is impossible to discover which failure occurred first.

The impact of concrete as it is placed in the forms sometimes is neglected. Most concrete specifications prohibit an appreciable free fall of concrete since this tends to produce segregation of the coarse and fine aggregate, but it is sometimes allowed when no provision has been made for it in the design of the forms.

The method of transporting concrete to its point of deposition can be a factor in the design of the forms. If motorized trucks or buggies are allowed on the forms for a slab, for example, the horizontal forces applied when these vehicles start or stop should be considered in designing bracing for the shores.

The collision of trucks, automobiles, cranes, and other vehicles with any part of the forming may cause an accident and very often cannot be anticipated. When the location of the construction site is such that collisions may be a problem, barricades and fences are a wise precaution.

Probably the two greatest causes of failure are poorly constructed joints and lack of bracing. Connections in general are likely to have too few nails or other connecting devices and, as often happens, nails are poorly driven and cause splitting of the wood. Bracing, like many connections, is often not really designed for strength. It is usually very indeterminate and has hard-to-define loads. The rules of thumb used in its design may or may not be adequate.

CHAPTER 13

Business Law Topics

13.1 TORTS

Any action that wrongfully injures a person or his property is a tort. The action may be intentional or unintentional, force may or may not be used, and the action may even be negative, as when failure to perform some act results in damage.

The distinction between a crime and a tort should be noted. If a wrongful act is an offense against a government and violates either common or statute law, a crime has been committed and the proper governmental police authority will bring suit against the offender. The less serious crimes are classified as misdemeanors and the perpetrators are usually punished by fines whose dollar amounts often have little connection with the monetary value of the injury. A serious crime is a felony and the penalty is often imprisonment or even death. Some torts are also crimes, but in most torts the injury is such that the injured party must bring suit to obtain justice. The penalty imposed on the offender is determined by the amount of the damages to the injured party and is paid directly to him.

Property torts are usually well defined and easy to measure since the cost of repairing damage to property or replacing it can be determined with some precision. Personal torts, however, lead to more trouble in the courtroom because of the impossibility of any precision in determining the value of damages to a person's body or his reputation.

13.2 NEGLIGENCE

Probably the most common type of tort committed by construction companies, engineers, and people in all walks of life is that of negligence. Causing harm to a person or property by failing to observe proper precautions is negligent conduct. The negligence may be intentional or not, and may consist of an act or the omission of an act.

In determining whether the proper precautions were taken or not, it must be shown that the offender did not do what a normally prudent and reasonable man would have done. Professional persons such as doctors and engineers are not expected to be perfect in all their actions, and testimony by similarly trained professionals is required to prove that they failed to perform in a normal way.

Malpractice suits against doctors, architects, and engineers are becoming more common today than ever before, and this type of negligence suit can be so damaging to the professional in terms of his reputation that many professional men find it necessary to carry insurance policies that will protect them against losses from this type of lawsuit. Many structural failures and construction accidents are caused by carelessness on the part of an architect or engineer, but it must be shown that the usual standard of care was not exercised before a negligence suit can be successful.

A tort may be either a private or a public offense. Causing damage to one or a few people is a private tort. Negligent design of a highway bridge, or failure to warn the public of an excavation in a highway, or inadequate lighting of roadway hazards could be the cause of injury to a large number of people or of the public in general, and should be classified as public torts.

Most negligence is of short duration or may be rather casual in nature, even though the results may be very damaging. A willful, reckless disregard for the rights of others may be considered gross negligence, or even criminal negligence if such recklessness leads to the loss of life or great physical damage.

Three types of action may result in some variety of tort. Malfeasance is an act that should not be permitted under most conditions and may result in a crime as well as a tort. Improper acts of public officials are often referred to as malfeasance in office and may be crimes as well as torts, and certainly are acts that should not be permitted under any conditions. Misfeasance is the performance of a lawful act in an unlawful manner. The discharging of firearms or fireworks may be a lawful act in certain remote or uninhabited areas, but the same activity carried out in a heavily populated district could easily injure others, and such misfeasance could be called a tort at least or probably even a crime. Nonfeasance, or the complete failure to act, is neither a tort nor a crime unless such failure to take action results in injury.

A person trained in first aid has no legal obligation to stop and render assistance at the scene of an accident; thus nonfeasance would not be a tort in this situation because no further injury would result from failure to act. If, however, such a person did attempt to aid an accident victim, he would be obliged to perform to the best of his ability. Rough handling of a victim with possible internal injuries or an injured back or neck could lead to a charge of negligence on the part of the "good Samaritan," even though there would be no penalty if he had completely ignored the injured person.

In determining guilt in a negligence suit it is often not clear precisely what caused the injury. There must be a definite sequence of events leading up to the accident or injury. The "substantial" or "proximate" cause of the accident is the one event without which there would have been no injury. A motorist may be blinded by the lights of oncoming traffic on a foggy night and fail to see the unlighted and unattended vehicle stopped in his lane of the road. If he failed to stop in time and collided with the stopped vehicle, there may be several factors contributing to the accident, but the direct or proximate cause was his failure to adjust his speed to the existing road and visibility condition.

When the facts surrounding an accident or other negligence case and the proximate cause can be established, there are two possible defenses that can be used against the injured party: contributory negligence and assumption of risk. Contributory negligence results when the injured party in some way contributed to his own accident. Carelessness on the part of a worker who violates established safety regulations may contribute to his own accident and make it impossible to collect damages. Certain jobs and professions are known to be hazardous and it is assumed that workmen in those jobs assume a certain risk as part of the jobs. A steeplejack's work is assumed to be dangerous, and if he is injured by falling while painting a tall tower because one of his supporting ropes broke, he cannot accuse the owner of the tower of negligence. If, however, the steeple he is painting is not structurally sound and the steeple topples and he falls, he may bring suit against the owner if this hazard was known and concealed from him.

The supplier or manufacturer of goods that might be dangerous to the user if improperly made or assembled must use reasonable care in making the goods safe if used as intended, or be guilty of negligence. This applies to structures and other engineering works as well as machinery and other manufactured products. Suppliers of foods and drugs as well as those involved in the sale and use of products that are inherently dangerous, such as explosives, are subject to "strict liability." They are liable for all injuries arising from the use of their products, whether used in the proper manner or not. Negligence need not be proved, and contributory negligence in their use is not a valid defense.

13.3 NUISANCE

Any act or condition that interferes unreasonably with the use or enjoyment of a person's land is a nuisance and the injured party may petition a court for an injunction, asking that the act or condition be discontinued, or he may ask the court to award him damages.

If the nuisance disturbs one person or only a few people it can be called a private nuisance. If the general public is concerned, a public nuisance results, and a criminal action rather than a tort may follow.

A nuisance is a long-standing condition, must be offensive or inconvenient, but need not be dangerous or injurious. Excessive smoke or odors, loud noises, and air and water pollution are examples. Obstruction of a highway by blasting or improperly flooding public property are examples of public nuisances.

13.4 TRESPASS

The act of entering or casting objects upon the land of another without express or implied permission is a trespass and the injuring party is a trespasser. Even though violation of one's property only once is a tort, a lawsuit should not be considered unless there was damage caused by the single trespass. Repeated trespassing is the usual cause of court action.

The landowner owes the trespasser no duty except to refrain from doing him bodily harm. Injuries sustained by the trespasser from natural hazards or other unsafe conditions are not the responsibility of the owner of the land. The owner may use force to remove the trespasser, but care must be taken that no more force than just enough to remove the trespasser is used.

The trespasser assumes the risk of possible injuries from natural or other causes in most cases. Injuries to children who are trespassing, however, sometimes become the responsibility of the landowner. The "attractive nuisance" doctrine has been in use in some states for about a hundred years, but its use is becoming more common. If a landowner has a device or property condition that would be attractive to children and at the same time dangerous, he must take precautions against injury to the children. The attractive nuisance is usually something unusual such as a fascinating piece of machinery or mechanical equipment, or an unfenced or unguarded swimming pool.

Social guests, sometimes door-to-door salesmen, delivery people, and firemen and policemen acting in the line of duty have an implied permission to enter a person's property and are known as "licensees." The landowner owes them the duty to warn of any unsafe conditions. Signs warning of an overly conscientious watchdog or verbal warnings of slippery pavement are necessary for the protection of the licensee, but the owner is not required to remove the unsafe condition.

When "invitees" enter an owner's property, they are there because they have been summoned or invited for business reasons or other matters of interest or profit to the owner. For invitees, conditions must be made safe, since these guests are entitled to a higher standard of care than are licensees. The slippery pavement must be made safe, the angry bull or watchdog must be restrained. Firemen and policemen summoned to the home or place of business are invitees and all hazards must be removed unless circumstances make it necessary for these guests to assume the risk of their professions.

13.5 DEFAMATION

The malicious injuring of the good name or reputation of another is defamation and may be either oral or written. A defamatory statement exposes another to public ridicule, contempt, or hatred, thus violating one of his property rights, the right to his earned reputation.

Verbal or oral defamation is "slander." A slanderous statement must be false and it must be shown that damage results from that statement. Whether false or not, it is assumed that damage results from some accusations, and slander exists in these cases:

1. Accusation of a crime or of moral turpitude,

2. accusation of suffering from a loathsome disease,
3. accusation of unchastity in a female, and
4. accusation of inability to perform a profession or business.

In these cases no damage need be proved, and truth is no defense.

Written defamation is "libel." The statement must be false, but, unlike slander, damage does not have to be proved since it is assumed that anything written and published will be read and believed by a third party or parties.

The best defense against a defamation suit is proof that the alleged defamatory statement is true, despite whatever malicious or evil motives prompted the statement. Under certain circumstances defamation is permitted. Courtroom proceedings and legislative hearings are "privileged," and judges and sworn witnesses have the right to defame others. A "qualified privilege" is given to certain columnists, critics, and others in the news media and, unless the privilege is abused, they also may defame with no fear of legal action.

13.6 INVASION OF PRIVACY

A person's appearance, name, and life history are his own private property and any violation of his right to keep them private, usually for commercial purposes, is a tort.

The publication without consent of pictures showing individuals in embarrassing situations is a violation of personal property rights or an invasion of privacy. Unlike defamation, truth is no defense, and no monetary damage must be proved.

Politicians, sports figures, entertainment celebrities, and others in public life are presumed to have waived their rights to privacy and any pictures of them, whether subjecting them to ridicule or not, can apparently be published with impunity.

13.7 ASSAULT AND BATTERY

The torts discussed so far have all dealt with offenses against or injuries to a person's property. Assault and battery are concerned with an individual's person, or his body.

The first of these, assault, is an intentional act that creates fear in the mind of the injured party of some bodily harm. Such a threat of violence must create the impression that harm is forthcoming immediately, not at some indefinite time in the future, and that the means of carrying out the threat are at hand.

Actually carrying out the threatened action is battery. Although the two actions usually go together, either assault or battery separately is possible. Both torts can also be classified as crimes.

The usual defense against these torts is a plea of self defense—whether it be defending his own person or property, or sometimes the defense of a third person or his property. The use of force, of course, must be reasonable, and unless care is taken that no more force than necessary is used, the personal rights of the attacker may be violated.

13.8 CONVERSION

The unauthorized use of the property of another is conversion. It may consist of the sale of property owned by another, destroying or altering property, or any other use of another's property without permission.

Using company funds for gambling and other speculative purposes and the use of company-owned equipment for private gain are typical examples of this tort. The line dividing stealing and conversion is not always clear, and although stealing is a crime, many conversion torts are also criminal.

The wronged party in conversion usually has some choice in the damages sought. Usually the market value of the goods in question plus interest is the amount of the damages, or if he prefers, he may act "in replevin" and seek return of the converted article rather than its monetary value.

13.9 PATENT

A patent is a right granted to a person by the U.S. government, which guarantees that person the exclusive right to make or use his invention for a period of 17 years. A patent can be thought of as a contract between the inventor and the government in which the inventor agrees to describe his invention to the public if, in return, the government agrees to protect the rights granted by the patent.

A patent must be issued to the inventor. A corporation, even though considered a legal person, cannot be issued a patent, although it may acquire the legal right to use it from the patent holder. Patents may be issued to two or more persons as joint inventors if they can prove joint efforts in the invention.

A utility patent may be issued for the invention or discovery of a new and useful art, machine, process, or composition of matter. An idea relating to any of these topics cannot be patented. Instead, the idea must be "reduced to practice," or made practical. Some of the classes of utility patents are:

1 A process, such as for separating a metal from its ore;

2 a machine, such as might be used to transform certain fuels into electric power;

3 a manufactured article made from raw materials, such as disposable food cartons; and

4 a composition of matter, such as a new clothing material.

A plant patent may be issued for the invention of a new variety of agricultural product or a new variety of ornamental plant.

Design patents have to do with the shape, appearance, or artistic design of an article of manufacture.

Responsibility for administering the patent law rests with the U.S. Patent Office in Washington, and applications for patent must be made through that office. Although the procedure for securing a patent is thoroughly prescribed by law, a patent attorney is often retained to help with the details, both technical and otherwise, of preparing the application. The application for a patent consists of three parts. The first part, the petition, is merely a formal request that a patent be issued. The second part, the specification, describes the invention in detail, and includes drawings as well as written explanations. Drawings must be in required form and are often drawn by specialists in patent drawing. The drawings plus the written explanation must be clear and complete enough that from them the invention could be built or otherwise put into use. If the invention is a machine, an explanation of the component parts and how they are expected to work is included. Included in the specifications are the claims, which state what has been invented and, thus, some idea of the protection expected from the patent. If the claims are too broad and indicate that a great deal of protection is expected, the patent application may be rejected. If the claims are too narrow, it may be possible to use some ideas in the patent without infringement. The advice of a patent attorney is often very useful in anticipating the possible uses of a patent and expressing the claims accordingly. The last part of the application is an oath in which the inventor declares himself to be the true inventor of the subject of the patent application.

The application must be checked by an examiner in the Patent Office to make sure that it complies with the patent law and that the invention is patentable. The investigation and search of the patent examiner is very detailed, and a year or more may elapse before the inventor is notified as to the result. In many cases some of the claims may not be allowed and the application rejected. The inventor may then revise his application and resubmit it if he so desires. The process of revision and reexamination may take place several times before the patent is finally granted or the application rejected by the Patent Office.

The fee paid by the applicant varies according to the number of claims made and also the time for which patent protection is requested. A 3½-year design patent, for example, requires a $10 fee, a 14-year patent $30, and so on. Patents other than for design require at least $30 as a filing fee, with higher fees accompanying larger numbers of claims.

The ownership of a patent must be in the name of the inventor, but if an engineer or other researcher has used his employer's time and facilities in creating his invention, the employer has "shop rights" and may use the invention without payment of royalties to the inventor. Many companies insist, as a conditon of employment, that any patent rights obtained be assigned to the employer. This would certainly apply to all inventions made using company facilities but may also extend to those made on the employee's time and using his own facilities if the agreement with the employer so states. There should be a time limit on such agreements, and very often the agreement to assign ends at termination of employment.

Infringement of patent rights is a tort, and enforcement of the patent rights rests with the patent holder. It is the responsibility of the inventor to make separate agreements with all who wish to use his invention, and he must take the initiative and bring suit against any who might use the invention without permission. Manufacturing companies very often unknowingly infringe upon patent rights. As many companies have discovered, it is much cheaper to require the inventor to protect his own rights than it is to conduct an exhaustive patent search, seek out the inventor, and negotiate licensing terms whenever a new process or idea is used. Claims made in a patent application are sometimes so vague that a court decision is required to determine whether infringement has been committed or not.

13.10 COPYRIGHT

A copyright gives an author the exclusive right to reproduce written works such as books, magazines, maps, photographs, music, motion pictures, and drawings for a period of 28 years, with a possible renewal for another 28 years.

Copyrights are supervised by the Library of Congress and are obtained by a relatively simple application, depositing two copies of the work with the Registrar of Copyright, and paying a small fee. Since no search is required, the procedure is less time consuming and much easier than that for a patent.

13.11 TRADEMARK

A trademark is used by a manufacturer or merchant to identify or distinguish his goods from similar products and may consist of a unique symbol, name, or word.

Applications for trademark are filed with the Patent Office and are investigated by an Examiner of Trademarks. The mark, when issued, is valid for 20 years and may be renewed as many times as required by its use in interstate commerce.

13.12 PROPERTY

Property, or anything owned, can be broadly classified into chattels and real property. Since real property is easy to define, chattels are anything that is not real property. Chattels personal are tangible, movable, personal prossessions. Chattels known as "choses in action" are intangible rights arising from tort action or contracts.

Real property consists of such fixed and imperishable items as land, buildings, trees, standing water or water below the surface, and any rights and privileges that accompany ownership of the land. Ownership of minerals beneath the surface, unless specifically excepted are owned by the landowner. Annual crops are considered personal property and do not belong to the land. The water from running streams may be used by the landowner but is not owned by him. Additions to buildings that are intended to become permanent parts of the building, such as a furnace or carpeting, are chattels until installed, then become fixtures or part of the building after installation and are owned by the landowner.

13.13 ESTATE

The title or interest that a person has in real property is called his estate, and there are several types possible. The highest type of ownership of land is called fee simple. It allows the owner complete use and enjoyment of the property as long as he does not infringe upon the rights of another. He may give or sell the land to anyone during his lifetime, and upon his death the land will go to his heirs according to the provisions of his will or according to the laws of the state if he has no will.

An estate for life gives full ownership of property for the life of a designated party. Upon his death, the ownership will revert to the original owner or someone designated by him. The owner of a life estate may sell or assign his right of ownership to another, but this right will terminate with his death. He may continue mining or farming operations, but opening new mines or exploiting the land in a new way would constitute waste and would be forbidden.

An estate for years is for a definite period of time. The rights and restrictions during that period are the same as for a life estate. The estate for years, although an interest in land, is classified as a chattel or personal property and upon death will be disposed of in accordance with the terms of the will of the estate holder.

13.14 OWNERSHIP

Ownership of land may be in the name of one party, in which case that party has sole ownership. If more than one party has an interest, there are several possible situations.

A tenancy in common results when two or more parties share in the ownership of land. Each tenant has an individual fractional interest in the land, and this interest may be sold, mortgaged, or assigned without the consent of the other owners. Each tenant has a part interest in the use, expenses, and profits that may arise from the property. The interest of a tenant in common will be disposed of according to the terms of his will upon his death.

A joint tenancy is similar to a tenancy in common except that upon the death of one of the tenants, his interest cannot be distributed to his heirs, but goes instead to the surviving joint tenant. If, during his lifetime, a joint tenant sells or assigns his interest to another party, a tenancy in common results and the joint tenancy is terminated.

A special case of joint tenancy, in which husband and wife together hold tenancy, is called a tenancy by the entirety. Neither party can sell or assign his interest separately. The death of one party will leave the survivor as sole owner. In case of divorce, the parties become tenants in common. Creditors of the husband or wife cannot attach the property to satisfy the debts of one individual party, but the property may be taken to satisfy a judgment against both of them.

Community property in some states refers to property acquired after marriage by husband and wife. Each has an equal share in it upon divorce, and upon the death of one the survivor becomes sole owner unless prohibited by the terms of the deceased's will. The survivor cannot be prohibited by the will from receiving at least half of the community property, however. Each party may have had separate property before marriage, and this property does not necessarily become community property after the marriage. Property may be acquired separately after marriage also without becoming community property.

13.15 DEED

A deed is the written and sometimes sealed document that transfers title or ownership of real property from one party to another. Laws pertaining to deeds are similar in most states, and in all states the deed must be registered in order for the transfer of ownership to be valid. The purpose of registration is to protect the public by making it possible to check the ownership of all land that might be offered for sale. Some cities and most county seats have a Registry of Deeds in which the deeds for all land transfers are recorded.

A typical deed consists of several parts. The grantor or person selling or giving up ownership of the land, and the grantee who is to receive the land must be clearly identified. The selling price or terms of sale must be made known. The land must be clearly identified, either by a verbal description, by a map or plat, or by reference to existing maps. Verbal descriptions may include names of abutters or may give distances and bearings of the property lines, starting from a well-defined point. Property lines may run from stakes, rock piles, outstanding trees, concrete posts, iron pipes, and may also run along highways, streams, and fences. The description should be clear and sufficiently defined that the location of the property can be established without question, now and at any time in the future. Any encumbrances existing on the land, such as unpaid taxes, water and sewer assessments, and all mortgages, should be specified and provision made for paying them. The type of deed is included in the words of conveyance. The entire deed must be signed, sometimes sealed, witnessed, delivered, and accepted by the grantee. Finally, the deed must be recorded as required by statute. If a deed is not recorded and the same land is sold again to another purchaser who does record his deed, the original purchaser could not prove ownership of the land and would lose all claim to it.

Depending upon the circumstances, there are several types of deed possible for conveying title to real property. The method of acquisition dictates the rights that accompany ownership of the land. A full covenant and warranty deed or, simply, a warranted deed is the best type of deed and gives fee simple to the grantee. It guarantees that the grantor has clear title to the land and has the right to convey it, that there are no encumbrances on the land not stated in the deed, and that the grantee and his heirs and assigns may forever have the right to peaceful enjoyment of the land. If the title to the land is ever disputed in the future, it is the responsibility of the grantor to take whatever action is indicated.

A quitclaim deed, as the name suggests, is one in which the grantor assigns any rights he has to the land in question to the grantee. He may not really have any valid claim, but he agrees to relinquish that claim, however large or small. Quitclaim deeds are often used in acquiring a warranty deed when certain relatives of a previous owner who may have died without a will agree to give up whatever small claim they may have had in the land. A quitclaim deed is also sometimes used as a first step in claiming property that apparently is not owned. Deeds are often carelessly written or land surveys are incorrectly made, with the result that nobody seems to own a certain piece of land, and a quitclaim from a party who seems to have no claim on the land will at least be a start in claiming ownership.

A bargain and sale deed or referee's deed may or may not be valuable. A bank or other mortgage holder, after foreclosing for nonpayment, usually sells the land, and grants a deed that simply transfers any rights the mortgage holder had to the grantee. Any defects in title arising from such a bargain and sale deed must be cleared up by the grantee. The deed itself describes exactly what rights the grantor has but contains no other guarantee or warranty.

13.16 ADVERSE POSSESSION

Property belonging to another may sometimes be claimed by adverse possession. If such property be used without permission, openly, against the true owner's interest, and continually for a period of twenty years or such time specified by the particular state, ownership of the land may be given to the trespasser.

A right-of-way is sometimes made legal by the process of adverse possession. If the public is allowed to use a road or walk across private property continually for the required twenty years and the owner has made no effort to prevent this use or to charge a fee for the privilege, at the end of that time the right-of-way may become public property. Property owners in this situation sometimes make it a practice to deny to the public the use of the way for one day per year, thus preventing the uninterrupted use required by the doctrine of adverse possession.

13.17 MORTGAGE

The purchase of land for a house or the purchase of an already built house represents a major investment for most people and usually requires that money be borrowed from a bank or other lending agency. The purchase of land by real estate developers for commercial uses must also often be financed by a bank. A mortgage results when the lending agency or mortgagee loans money to the mortgagor and in the resulting contract the mortgagor promises to repay the money along with a stated amount of interest. If he fails to repay the loan as required, the mortgage gives the lender of the money the right, if necessary, to force sale of the property in order to receive payment. The mortgagee thus has a lien on the property. Since a mortgage is an interest in real property, it must be in writing and must also be recorded so that any interested may be aware of this encumbrance on the land.

When there is a mortgage on property, the mortgagor does not have unlimited rights in his use of the land. Since the property is security for a loan, waste by the mortgagor cannot be tolerated. The loss of buildings and other fixtures on the land might make the property worthless and the mortgagee usually insists on full insurance coverage. The complete exploitation of mineral rights might also reduce the value of the property to a point less than the value of the loan, and the mortgage should prohibit this practice. The mortgagor may sell or will the property to another and the grantee must also accept the terms of the mortgage. The mortgagee may also sell or otherwise assign the mortgage without the consent of the mortgagor.

The mortgagor is usually considered to own mortgaged property subject to the claims of the mortgagee. State laws state the conditions under which the mortgagor may lose title, or how many payments he may miss before the mortgagee may force the sale of the property. If property is sold, court costs, then the mortgage balance is paid, then any second mortgage or other lien, and finally any remainder is returned to the mortgagor. If the sale of the property fails to yield enough to pay the balance of the mortgage, it is felt that the mortgagee, in taking a business risk, evaluated the property too highly, and any loss involved is his. He is not entitled to seize any other property of the mortgagor, since only the land in question was meant to be security for the loan.

13.18 TITLE SEARCH AND SURVEY

Before title to land is conveyed, a title search should be made. The purpose of the search is to ensure that the grantor of the deed has clear title. By checking all previous deeds it should be possible to follow a complete and unbroken chain back far enough in time that there is no possibility of other claims to the land arising in the future. In addition to the unbroken chain of clear ownership, the search should uncover mortgages, taxes, special assessments, court judgments, and any other attachments that might cloud the title.

It is often necessary to have the land surveyed before a proper deed description can be written. The purpose of the survey is to locate the original boundary lines on the ground. Distances and bearings contained in old deeds are often incorrect, with the actual distances being larger than recorded in the deed. Magnetic bearings are often used, and the surveyor must determine the magnetic declination at the time the survey was made in order to trace boundary lines. When streams and rivers are used as boundary lines, any changes in the course of the streams must be known, since the position of the stream when the survey was made determines the boundary line. Highways are also sometimes changed, and their positions when the original survey was made must be known. When corner markers are missing as they so often are, and the distances and bearings are also wrong, determining the true position of property corners becomes a difficult matter for the surveyor and demands a great deal of judgment and tact. When the actual boundaries and the area given in the deed do not agree, the boundaries control. When the deed description and the actual boundaries do not agree, the boundaries again control.

13.19 EASEMENT

An easement is an interest in land and is the right to make use of the land of another for a definite and specified purpose. Electric transmission lines, pipelines, and driveways and other access roads frequently cross private property, and easements are granted for that one purpose only. The easement is conveyed by deed, can be sold or assigned, can be inherited, and can be transferred independently of the land. When the need for the easement ceases to exist, the easement also ceases to exist. Thus, if an access road to previously inaccessible property is no longer necessary or used, the easement for that road will also no longer be needed. An easement may be terminated at any time by mutual consent.

13.20 LICENSE

A license is permission to do what might otherwise be trespassing. It may be oral rather than written, may be revoked at any time, may not be assigned, need not be recorded, and is simply an unenforceable privilege, not an interest in land. Its purpose may be the same as an easement, but its terms are often of such short duration that the formalities of an easement are not necessary.

13.21 FRANCHISE

A franchise is the right granted by a government—often a city or town—to operate a public utility over land owned or controlled by that government. A franchise is personal rather than real property.

13.22 EMINENT DOMAIN

The right of a government—federal, state, or local—to take private land for public use is known as eminent domain and is considered to be a higher right than private ownership.

The power or right of eminent domain is usually exercised by a government in acquiring the right-of-way necessary for highways, dams, canals, or other public utilities, as well as schools, hospitals, and other public improvements. The power may also be delegated to private businesses that serve the public, such as electric companies, railroads, and fuel-transmission-line owners.

Property taken by eminent domain must be paid for at a reasonable market price, and the price offered may be appealed to a court. Only as much land as is necessary for the public improvement may be taken. The taking of land by eminent domain or condemnation and procedures used in solving disputes are usually well defined by state or local statutes.

13.23 WATER RIGHTS

Boundaries are often determined by bodies of water and the boundary location and rights to the use of the water are governed by the type of water and appropriate local law and court rulings. With the passage of time and the growth of the population, water rights are becoming more important and more varied in the several states. Many water rights come from English common law, but recent legislative action and court decisions have complicated the matter greatly, so there are now many differences in water rights, corresponding to local conditions.

13.24
SMALL STREAMS AND LAKES

Water on the surface of the earth flowing in an established bed or channel and having a regular source of supply is referred to as a water course. A lake is a nonflowing body of water located in a natural or artificial depression in the earth's surface. Streams and lakes that are too small to be used by boats in regular commercial travel are held to be nonnavigable and may be owned entirely by a private party, or the boundary of a person's property may be such a stream or lake. One whose property borders a stream or lake is a riparian owner and, as such, he has certain property or riparian rights concerning his use of the water.

When a stream forms the boundary between two pieces of land, each parcel of land is considered to extend to the center of the stream, or "the thread of the stream" when the stream is at its lowest stage. Sudden changes in the course of the stream do not change the boundaries; the center of the old stream bed remains the property line. The path of a slow-moving stream, however, constantly changes and material removed from one bank and deposited elsewhere can cause a property owner to lose or gain land as the process of erosion and accretion continues.

Ownership of small lakes extends to the center of the lake unless the deed stipulates otherwise. In some states lakes larger than a certain acreage are considered public property, in which case private ownership extends only to the shoreline.

A riparian owner has the right to reasonable use of the water adjacent to his property. He expects the water that comes to him from upstream to be undiminished in quantity and quality, and he owes property owners downstream from him the same duty. When the quantity of water in a stream is limited, domestic use for drinking, bathing, and washing has priority over agricultural use. The use of water for irrigation has a lower priority than water needed for the care of livestock. It is wrong to deprive a downstream neighbor of his usual quantity of water, but it is equally wrong to construct a dam and flood the land or cause water to enter the basement of an upstream property owner, and such a situation would make the dam owner liable for damages.

In some of the western states large quantities of water are used for irrigation and sometimes water is brought great distances for this purpose. The use of water in these states is sometimes governed by the principle of prior appropriation, or simply appropriation. This principle allows the first person to use water to continue it in such quantities as are deemed necessary for his purposes. The result of this appropriation may, of course, leave no water in the stream for the use of property owners downstream. The right of appropriation is quite different from ordinary riparian rights, in which the purpose is to share the use of the scarce water among all the riparian owners. With increasing demand for water, some states are modifying the old appropriation rights somewhat in the direction of ordinary riparian rights.

The right of appropriation is sometimes justified by applying prescription rights which allow long-standing conditions to continue. Thus, if a farmer has appropriated nearly all the water in a stream for irrigation over a period of many years, his prescription rights will allow him to continue to do so. Also, if a manufacturer has been allowed to pollute a stream for many years, his prescription rights will protect his right to continue this practice. The riparian rights of owners downstream from the manufacturer prohibit him from making matters worse but can do nothing to force him to improve the situation.

13.25 GROUNDWATER

Water below the surface may flow in a regular pattern, in which case it becomes a subterranean stream and the rights to the use of the water are the same as surface riparian rights.

Most water below the surface does not flow in a regular channel and is classified as percolating water. The principal sources of percolating water are rain and melted snow, which penetrate the soil, and water from streams and lakes, which slowly permeate the soil in all directions. The percolating water moves but in no well-defined pattern and is considered part of the land overlying it. Wells may be drilled or dug to tap an underground source, and the water may be used without limit unless it can be proved that an unreasonable use so lowers the water table that other property owners suffer. Either lowering the water table or raising it, thus causing injury, may result in a suit for damages.

13.26 NAVIGABLE WATERWAYS

Waterways large enough to maintain commercial boat traffic are designated as navigable waterways and are the responsibility of the federal government. Property that borders upon navigable water can be privately owned to the mean low-water mark, but the owner has the right to install a pier to the area of navigability. Where the navigable water is affected by tides, private ownership extends only to the high-water mark; the land between high and low tide is owned by the state. Since mean high- and low-tide levels are difficult to determine with any precision, some states have defined property and riparian rights by more exact means.

Many state boundaries are navigable streams, and definite rules establishing ownership have been the source of much legislation and court action. Some states own to the thread of the bounding stream, some to the near shore, and some to the far shore.

Control and regulation of navigable waterways are given to the U.S. government, and dredging of harbors and channels, maintenance of navigation aids, and decisions determining navigation lines and clearances are usually delegated to the Army Corps of Engineers.

The construction of reservoirs, sewage treatment plants, power plants, and industrial plants that may cause a change in the quantity and quality—including temperature—of the water is regulated by the Corps of Engineers, and their requirements must be met before starting any project using water under their control.

CHAPTER 14

Construction Legislation

There are at least two widely held concepts of the construction contractor in action. At one extreme he is pictured as the last of the independent free-wheeling entrepreneurs, hiring and firing workmen at will, making outrageous profits, controlling local and national political figures, and, in general, emulating the activities of the robber barons of the last century. The other extreme point of view represents the contractor as being so repressed by governmental red tape and union restrictions that he no longer is the manager of his own company, unable to make an independent decision. The truth, of course, lies between these two extremes, but since his actions are somewhat restricted because of governmental statute, some of the recent legislation relating to the construction industry should be studied.

14.1 WORKMEN'S COMPENSATION LAWS

Until the early years of the twientieth century it was assumed that industrial accidents, causing death or injury, were the responsibility of the workmen, and that they assumed the risk of injury as a condition of employment. The employee could obtain compensation for his injury if, in a lawsuit, he could prove the employer negligent, prove himself not negligent, and prove that the accident was caused by a fellow worker. This procedure required a fellow employee to jeopardize his future employment possibilities by testifying against the employer. This situation, plus the time and legal expense of a lawsuit, made it obvious that legislative action must be taken.

Starting in 1908, all the states have enacted workmen's compensation acts which are similar in nature. Unless it can be proved that the injured party was intoxicated or showed wilful intent to injure himself, he is entitled to compensation benefits for any injury. The laws are usually administerd by a state board or commission, but appeals may be made to a state court.

In the case of temporary disability the benefits usually start after one week and may amount to between one half and two thirds of the injured worker's normal weekly pay. In cases of permanent disability a lump-sum payment is sometimes substituted for payments over a long period of time. Benefits over a definite period or a lump-sum payment are made to the heirs or dependents of the worker killed in an accident.

Most of the states require the employer to carry sufficient insurance to cover all possible injury claims. In some states the employer simply contributes to a state fund; in others he may purchase insurance from a private company, or he may even be qualified to act as his own insurer.

14.2 DAVIS–BACON ACT

The Davis–Bacon Act, passed by Congress in 1931, requires that workmen on federal-financed construction with a value exceeding $2,000 be paid wages not less than those established by the Secretary of Labor as the prevailing wages of the area. These minimum wages are made part of the contract for each specific project and are usually equal to the minimums established by local union contracts.

14.3 BUY AMERICAN ACT

The Buy American Act, first passed by Congress in 1933, is designed to protect American manufacturers from low-priced or unfair foreign competition. In general, if a raw material or manufactured article produced by an American supplier is available in sufficient quantity and quality, it must be used in preference to an equal product from outside the United States. Exceptions to this rule may be permitted by the government as changing foreign policy dictates. Some states have followed this policy further by insisting that materials used in state public works projects be furnished by in-state suppliers whenever possible.

14.4 COPELAND ACT

The Copeland Act, passed in 1934, makes it illegal for an employer to withhold pay due a worker on federal construction by force or threat of force or any other form of coercion. Union dues may be deducted only with the consent of the worker and then only if it is provided for in a proper collective-bargaining agreement. Periodic sworn statements are required to prove that workers are paid wages as determined by the Secretary of Labor.

When this act was passed in the depths of the Depression of the 1930s, employment of any sort was scarce and many workers were willing to refund part of their pay to the employer simply for the privilege of having a job. The obvious purpose of the Copeland Act was to prevent these kickbacks.

14.5 WALSH–HEALEY ACT

The purpose of the Walsh–Healey Act, passed by Congress in 1936, was to ensure that nonunion employers would have no advantage over union employers in bidding on contracts for supplying materials or goods to any federal agency. Minimum wage rates for various classes of workmen as determined by the Secretary of Labor were specified and overtime and overtime pay were defined. Child labor was outlawed and a minimum age of 16 for most labor or 18 years for hazardous occupations was stipulated.

14.6 FAIR LABOR STANDARDS ACT

The purpose of the Fair Labor Standards Act, passed by Congress in 1938, in the late years of the Depression, was to stimulate business by increasing the number of employed workers. Since it applied to industries in any way associated with interstate commerce (with some exceptions), its provisions covered many workers who did not come under the provisions of the Walsh–Healey Act.

Overtime was defined as any work in excess of 8 hours in one day or 40 hours in one week. An absolute minimum wage, not defined by a specific craft or trade, was established but has been raised several times since. The Walsh–Healey minimum wages were extended to cover many more industries.

14.7 NATIONAL ENVIRONMENTAL POLICY ACT

Problems associated with eliminating air and water pollution and preventing further deterioration of the environment are responsible for the formation of many private organizations which, through court action and lobbying in Congress, are forcing many developers and designers to justify their plans or abandon them. Partly as a result of their activities, much legislation has recently been passed relating to preservation of the ecology, and many governmental agencies have some, often overlapping, responsibility in environmental matters.

Of interest to the design profession is the National Environmental Policy Act of 1969. One requirement of this act states that an Environment Impact Statement must be prepared for any construction in which federal grants or loans of money or permits of any sort are required.

The impact statement is usually prepared or at least furnished to the responsible federal agency by the designer and must show the effect of the proposed construction on such factors as land use, zoning, groundwater and geology, surface drainage and erosion, recreation, wildlife, real estate values, agriculture, and social and psychological effects on the population. Not only such obvious items as air and water pollution, but noise pollution, long-range problems of streamflow, thermal pollution of streams, and many other factors must be considered.

Approval for a project is first granted by the federal agency involved, which then forwards the impact statement, or EIS, to the Environmental Protection Agency, or EPA. The EPA must review, in addition to the EIS, any comments and criticism it may have received from other governmental agencies or private organizations.

If the EPA rules the EIS as unacceptable, it refers the matter to the President's Council on Environmental Quality, which may then recommend to the President that the project be either allowed or halted. The President may or may not follow the recommendations of the Council.

Alternative proposals should be included in the impact statement as well as their effects upon environmental conditions in case the primary proposal is deemed unacceptable for some reason.

14.8 OCCUPATIONAL SAFETY AND HEALTH ACT

The safety record of the construction industry is poor, with nearly 3,000 deaths and more than 200,000 disabling injuries being reported each year. The Occupational Safety and Health Act or OSHA as it is referred to, was passed by Congress in 1970 in the expectation of reducing accidents.

The act covers all construction work associated with interstate commerce and applies to subcontractors as well as to general contractors and to engineering and architectural designers. The law simply requires every employer to provide employment which is free of recognized hazards that are likely to cause death or serious physical harm, and also to comply with OSHA standards. The administration of the law is to be by the Secretary of Labor and authorizes his representatives to conduct surprise inspection visits to all construction sites when it is believed that dangerous conditions exist. The inspector may question any employee privately and may require testimony under oath. Citations are posted for all violations of safety standards and penalties up to $20,000 and one year in prison may be levied.

OSHA standards or regulations cover all phases of construction activity, including such items as protective equipment; fire protection and prevention; signs, signals, and barricades; materials handling and storage; power tools, and blasting and explosives.

Many aspects of the cost of complying with the law are as yet unknown, as are the effects of the law on the liability of designers to contractors and owners when their designs may induce violations of the law by the builders. The cost of manufacturing new equipment and modifying old equipment in line with the new and constantly changing standards is as yet open to conjecture.

The bill, as originally passed by Congress, authorizes and encourages the individual states to develop plans of their own. When a state plan and its method of enforcement have been approved by the Department of Labor, federal standards will no longer apply and the state plan will be followed and enforced. Encouragement in the form of promised financial aid has stimulated many states in planning OSHA-type legislation.

14.9 LABOR UNIONS

The legislation referred to so far has dealt exclusively with the relations between employers and employees as individuals and relations among government agencies and designers, contractors, and owners.

The construction industry today is highly organized into craft unions and the workers are paid high wages, have substantial fringe benefits, and, except for some local problems, usually enjoy a high rate of employment. Conditions were not always thus, of course, and many years of struggle were required to achieve these conditions. The unions have been responsible for many of the gains made by construction workers, as well as all industrial workers, and some of the Congressional acts that have contributed to their power should be noted.

14.10 SHERMAN ANTITRUST ACT

The history of labor unions in this country during the ninteenth century shows slight gains in wages and working conditions in the face of hostile employers and an unsympathetic public and government. Violence by both parties to a dispute was common, and even after the primary weapon of unions, the strike, was finally declared legal, employers were often able to get courts to grant injunctions against strikes, thus making the unions' most effective efforts once again illegal. In the eyes of much of the public, unions continued to be thought of as criminal conspiracies.

The purpose of the Sherman Antitrust Act of 1890 was to curtail the growth of business cartels and prevent the formation of monopolies. In the eyes of many, union activity was aimed at a restraint of trade, and some sympathetic courts used the powers granted by this act to issue injunctions and thus stop both strikes and boycotts by unions. The Supreme Court, furthermore, ruled that both a union and its members could be sued for damages as a result of a boycott sponsored by the union.

At the beginning of the twentieth century, unions still had little power, a generally poor reputation, and little sympathy in Congress. What little legislation there was acted to restrict union activity. Legally the unions had few rights and many restrictions and in their bargaining the advantages were all on the side of the employers.

14.11 NORRIS–LAGUARDIA ACT

The attitude of the general public and Congress toward unions gradually changed with the excess of power exercised by employers as a result of the Supreme Court decisions interpreting the Sherman Act until Congress finally, in 1932, passed the Norris–LaGuardia Act. The principal result of this act was to protect the rights of workers to strike and picket peacefully without fear of injunction by the federal courts.

During the Depression of the 1930s employment of any sort was desperately sought, and many workers were willing to agree not to join a union as a condition of employment. These "yellow-dog contracts" were also declared illegal by the Norris–LaGuardia Act.

14.12 WAGNER ACT

The power of organized labor gradually increased during the early 1930s, but the Wagner Act, passed by Congress in 1935, gave the unions their greatest boost. It forbade employers from interfering in any way with union efforts to organize workers and forced employers to bargain with properly chosen union representatives. Unfair labor practices were defined and the National Labor Relations Board was given the power to administer the law and act on complaints from the unions against employers.

Unfortunately, the Wagner Act gave much power to the unions but established no restraints on this power. The resulting actions by organized labor once again turned public opinion aganist unions, as enormous economic power was vested in a very few industrial labor leaders. Actions by a few large unions during World War II that were contrary to public welfare and criminal activities by a few leaders during and after the war led the way to the next step in the rise and fall of union power.

14.13 TAFT–HARTLEY ACT

In an effort to force both unions and employers to act in a more responsible manner, Congress in 1947 passed the Taft–Hartley Act. Unfair labor practices on both sides were defined and machinery to handle disputes was set up. Although passage of this act was vigorously opposed by labor organizations, and many dire predictions were made concerning its effect on the unions, in the eyes of the general public the law has had a beneficial effect on labor relations in general.

According to the terms of this act the employer was forbidden from interfering with the rights of employees to organize or join unions, to interfere in any way with union activities, or to refuse to bargain collectively with a properly chosen union. These provisions were very similar to those of the Wagner Act but were perhaps spelled out more precisely.

Unfair acts by labor were defined by Taft–Hartley and were based largely on some of recent union activites. Unions were prohibited from encouraging membership or activity in their organization either by force or by applying pressure on the employer, collective bargaining in good faith was made mandatory, secondary boycotts, sympathy strikes, and jurisdictional strikes were outlawed. In addition, excessive union initiation fees, union contributions to campaigns for federal office, and featherbedding practices by unions were forbidden.

Enforcement of the law is vested in the National Labor Relations Board. Charges of unfair action can be brought by employer, employee, or the union itself, and after investigation, the regional NLRB must recommend punitive or other action. Appeals to the national NLRB or to the federal courts are provided for. Unions that request the services of the NLRB are required to disclose details of their organization and finances, and their officers must submit noncommunist affidavits.

Damage suits by either party to a dispute are theoretically possible under the terms of this act. Breach of contract suits or suits for damages because of wildcat strikes or secondary boycotts, for example, are now possible. Until passage of this act, action in the courts by unions against an employer for unfair acts was common, but it was very difficult for the employer to get any relief from the unfair acts of the unions. This new union responsibility is still difficult to enforce because of the difficulty of identifying the particular union officer or agent responsible for the unfair or unlawful acts.

14.14 LABOR–MANAGEMENT REPORTING AND DISCLOSURE ACT

The Taft–Hartley Act of 1947 instituted many needed reforms in the relations between unions and employers, but many changes in these relations were still needed. In addition, some unions were controlled by unscrupulous leaders, and the Reporting and Disclosure Act of 1959 was designed to correct some of the internal affairs of the unions as well as amend the original Taft–Hartley provisions. The authority for this act comes from the power of Congress to control interstate commerce, and all labor and management groups that have any connection with interstate commerce are under the jurisdiction of the act.

Under the provisions of this act unions are required to file with the Secretary of Labor details on union organization, elections, and finances whether they use the services of NLRB or not, and former Communists are prohibited from holding high union office.

Under Taft–Hartley such practices as secondary boycotts and "hot cargo" actions, although prohibited, continued, but not with the obvious sponsorship of the union. The new act tried to prevent such actions by individuals within a union or by separate employers.

Prevention of extortion by labor leaders against employers is the purpose of a clause defining payments that can be made to workers. Such fringe benefits as vacation and severance pay and contributions to union-sponsored training programs are still allowed, however.

Rights of a union in negotiating a labor agreement even though a majority of workers are not members and many articles pertaining to closed-shop regulations, clauses pertaining to required training and experience, and clauses concerned with hiring-hall practices by unions are included in this act.

The Taft–Hartley and Reporting and Disclosure Acts have done more than any other legislation in correcting some of the evils perpetrated by both labor and management groups against each other and against the general public as well. Much, of course, remains to be done, and more restrictive legislation will be forthcoming as the injured public demands protection against the results of the labor–management conflict.

CHAPTER 15

Litigation

It is not necessary that each engineer or contractor have a complete knowledge of the law and legal precedures, but some idea of legal practice is very useful to all professional men. There are so many occasions in which an engineer or architect or contractor must secure the services of an attorney or must deal in some way with an attorney that some idea of how the law operates is a necessity.

Most engineering work involves a contract of some sort, and contract disputes sometimes lead to courtroom appearances. Any professional man dealing with employment of others may find himself appearing before a court of law or official hearing board. Any professional whose expertise is in a field of public interest may be called upon to advise legislative or investigative bodies. Anyone with technical training or experience of any sort may have the opportunity to appear in court as an expert witness.

Courtroom actions are usually very formal and definite rules of procedure must be followed. Hearings before legislative bodies, on the other hand, may be much less formal, but the rules followed by most bodies are at least based on ordinary courtroom custom, and these customs should be understood by all competent engineers and architects.

15.1 FEDERAL COURT SYSTEM

Disputes involving the U.S. Constitution and cases based on the violation of federal statutes are tried in appropriate federal courts. Any case involving disputes between states, or cases involving ambassadors and other high public officials, important cases between citizens of different states, or disputes between U.S. citizens and those of a different country are typical of the workload of our federal courts.

The lowest court in the system, the District Court, is the most numerous, and the districts are arranged by Congress so that each state has at least one, with more than 100 now in existence. District Courts are courts of original jurisdiction, and most federal court cases have their first hearings in these courts. Special courts have also been established to deal with tax claims, customs matters, and money claims arising from contracts or other disputes with the federal government.

For each federal circuit there is a Circuit Court of Appeals presided over by three judges. These courts handle appeals from the lower courts and act on matters of interpretation of law as opposed to matters of fact, which are decided in the District Courts. Circuit Court decisions are usually final, but occasionally these courts may request a further review, or sometimes the Supreme Court may demand that a case be sent to them for review.

The highest court in the land is the U.S. Supreme Court, which consists of nine judges. This court has original jurisdiction in cases involving the Constitution and in disputes in which high federal officials are participants. Most of its activity is in cases appealed from the Circuit Courts and reviewing some cases from the highest state courts in which one of the states is a party. The Supreme Court's existence and duties are prescribed by the Constitution, whereas all other federal courts have been established by Congress as the need for them arose.

15.2 STATE COURTS

The names, duties, and descriptions of the state courts vary from state to state, but most state courts are patterned after the federal court system. They may be established by the constitution of the state, or they may have been set up by legislative action.

The lowest courts have original jurisdiction over such matters as probate of wills, juvenile matters, small claims, traffic violations, and various misdemeanors.

The next higher court usually covers a large geographical area and has original jurisdiction in disputes involving a relatively large amount of money, handles some felony cases, and may hear appeals from a lower court.

The highest state court is similar to the U.S. Supreme Court and hears appeals from lower courts. It is usually referred to as the state supreme court.

Some states have intermediate superior courts below the level of the supreme court whose purpose is to hear as original cases some of the more serious felonies and some of the most important civil cases. They may also act as appeals courts in an attempt to save time and lower the work load of the highest court of the state.

15.3 COURTS OF EQUITY

The legal system of this country is based on the system developed through the centuries in England, and most of our laws come from three sources. Constitutional law is spelled out in the U.S. Constitution and the constitutions of the various states. The legislative bodies—the U.S. Congress and the state legislatures—are constantly passing new laws and repealing others, and their regulations are referred to as statute law. Another type of law, the law governing most business dealings, comes from the accepted way of doing things and is known as common law. Common law comes from court decisions of the past rather than from laws passed by any governing body and has gradually accumulated and been compiled to the point where there are rules of conduct governing most phases of life. In spite of the vast accumulation of laws and recorded decisions, however, there are many situations that demand action not specified by any existing law. The fair treatment or justice required here is referred to as equity.

In England and formerly in this country there were two parallel systems of courts, courts of common law and courts of equity. The function of the law courts was to determine the facts in a dispute, determine which law applied, and reach a decision on the course of action to be followed. In a court of equity there is a jury only if the judge requests it to act on matters of fact, and the judge must rule on questions of law. Since no state statute or common law has been broken, the function of the judge is to determine what is fair to both parties. His resulting decree simply instructs one party to the dispute to perform or refrain from a certain action. Punitive action seldom results, and usually an injunction is issued or one party may be instructed to compensate the other party for damages.

Formerly the two types of court were separate and sometimes different aspects of one case were handled by different courts, perhaps in different locations and with different judges. The distinction between law and equity courts has now been removed in most states, with the result that a single court can handle both types of case and the law and equity questions in a lawsuit can be resolved concurrently.

15.4
TRIAL PROCEDURE

In a criminal case the responsibility for initiating action rests with the state, whose law has been violated, in the person of the district or county attorney. In a civil action, however, the injured party—plaintiff or complainant—must take the first step against the opposing party or defendant.

Rules vary somewhat in different localities, but the first step in a civil action suit is often a summons, which is sometimes issued by the court and sometimes by the attorney for the plaintiff. The summons is simply a notice to the defendant that legal action against him is being instituted.

After the suit has formally started, pleading by both parties takes place. Pleading by the plaintiff takes the form of a complaint, declaration, or petition in which his position is stated. In it all the facts of the case are stated, the reasons for seeking court aid, and the remedy hoped for.

After receiving the complaint, the defendant must answer within a stipulated time. Failure to respond with a plea of his own results in a judgment in favor of the plaintiff by default. The answering plea takes the form of one of three possibilities. A demurrer admits that the facts stated in the complaint are correct but argues that these facts are not sufficient for the legal action requested. A motion to dismiss the case should be based on fact; discrepancies in the facts presented by the petition must be shown. The third possibility is a counterclaim in which the facts stated in the petition are admitted, but other facts are introduced which show that damages are due the defendant rather than the plaintiff. In the case of a counterclaim, the plaintiff's reply or plea is called replication. The purpose of the pleading is to establish the grounds for the trial. The questions of either fact or law that must be decided by the trial must be clearly defined.

The jury in most civil cases consists of 12 members, although smaller juries for some cases are permitted in some states. Jury members are chosen from a list of prospective jurors with the agreement of attorneys for both sides. Prospective jurors may be disqualified at the request of one side if there is a possible conflict of interests. In addition, each side is usually allowed to disqualify a small number for arbitrary reasons.

Although trial proceedings are often very lengthy, the sequence of events in both criminal and civil cases is about the same. After the jury has been sworn in, or impaneled, each party's attorney briefly states to the jury what he is going to prove and, in general, how he is going to do it. The evidence is then presented by witnesses for each side, followed by closing statements from each side in which pertinent testimony is emphasized or amplified to make the strongest possible case. The judge then summarizes the case and instructs the jury as to its duties, usually reminding them that the court must decide points of law, whereas the jury is concerned with the truth of the facts presented. After the jury reaches its verdict, the judgment or official decision of the court is announced.

The procedure for enforcing the judgment may be very simple, such as a court order authorizing an official to receive payment of the stipulated damages or penalty. If a simple payment is impossible because of the large amount a levy may be issued by the court, which authorizes seizure and sale of property. The property is sold at public auction and any amount realized from the sale in excess of the judgment must be returned to the owner.

The losing party in a lawsuit may request a new trial if errors in procedure in the first trial or errors of fact may be shown. Faulty decisions by the judge or conduct allowed by the judge frequently are the reasons cited for requests for a new trial. Either party, however, may request a review of the case by a higher court. The reasons given must show an error in the application of law, since facts as determined by the lower court are presumed to be correct.

15.5 EXAMINATION OF WITNESSES

The question of fact or exactly what happened is determined by questioning witnesses, who are ordered by a subpoena to appear and give the required testimony. Failure or refusal to appear is an action in contempt of court and can be punished by imprisonment or fine. It is the responsibility of the plaintiff to prove conclusively that the defendant is guilty, and this proof must come from the evidence presented by the witnesses.

Each witness is questioned, under oath, by the attorney for the side that called him. Objections to his testimony may be raised by the opposing side and the presiding judge must then rule on whether to accept or reject the testimony or any part of it. After the first or direct examination, the opposition tries to break down the credibility of the witness or refute his testimony in some way by his cross-examination. As before, objections may be made and ruled upon by the court.

The rules of evidence concerning what is permitted are very strict but vary somewhat with the type of court and the jurisdiction or locality. The categories to be discussed here and the meanings of the terms are common to both criminal and civil cases and should be understood by any prospective witness.

Information presented by witnesses is sometimes characterized as being either real evidence or testimony. Real evidence consists of tangible objects such as the murder weapon in a criminal case or a broken cable in a civil accident case. Testimony is nontangible evidence such as verbal accounts or descriptions.

Much authority concerning evidence and its admissibility is in the hands of the judge. Certain facts that are common knowledge need not be proved and the judge will take judicial notice of them. All evidence must have a direct bearing on the case at hand and must be presented by a person who has a legal right to be present (that is, a competent person). Extraneous material may be rejected as being either irrelevant or immaterial.

When the contents of a written document or paper, such as a contract or deed to land, are important in a case, the document itself should be produced according to the "best evidence rule." If the document exists but it is impossible to produce it, secondary evidence may be substituted for it. If the deed to land cannot be produced, evidence of mortgage payments on the land may be the only evidence of ownership available and under some circumstances might be accepted.

There are many kinds of verbal evidence possible, not all of which are admitted. Direct evidence presents the facts of a case by relating what the witness actually saw happen, or the happenings as he experienced them. Direct evidence, if properly presented, is always acceptable since it is a very efficient way of showing exactly what has occurred. Circumstantial evidence, on the other hand, requires the jury or judge to infer a certain happening from given factual, external data. Since the inference made from circumstances surrounding the case requires some reasoning ability, and since that ability of members of the jury is unknown, circumstantial evidence is usually not permitted. There is the possibility also that the facts as related might lead to other than the desired conclusion, so that even if permitted, the use of circumstantial evidence is somewhat risky.

The use of direct evidence permits the witness to rely on his senses of sight, feeling, smell, and, to some extent, hearing. Identifying some sounds heard such as breaking timbers or gunshots is the basis of acceptable evidence, but a verbatim recounting of what someone said is classified as hearsay and is usually inadmissible as evidence. Courts are wary of hearsay because the original testimony was given when the speaker was not under oath, and also because there is no opportunity for cross-examination. There are some situations in which it is generally believed that a person will tell the truth and thus a statement by a person who believes that he is dying, although hearsay, may be admitted in court. Spontaneous utterances and declarations of pain, given by accident victims, for example, are usually made by the victim without intending to fool anyone and are thus allowed to be repeated. Declarations or confessions that act against the best interests of a person and book entries made in the regular course of business are also considered reliable and may be accepted.

15.6 OPINION EVIDENCE

Although it is not desirable, much opinion evidence invariably is introduced into a trial. Even that which appears to be direct evidence may inadvertently contain some opinions of the witness. Since opinions are based on the reasoning of the witness, most witnesses are allowed to express opinions on subjects that do not require any special skill or knowledge. The layman is permitted opinions concerning such things as estimates of number, color, such items as height and weight of people, temperature, and similar topics connected with everyday experiences.

Expert witnesses are permitted to express opinions on subjects in which they have some special skill or knowledge. It is assumed that the expert on any specialized subject has acquired, through study or experience, the ability to draw a correct conclusion from the given facts, or the gathering and presentation of the facts used to support his conclusion may be left completely up to him. Before such testimony can be permitted, however, it must be shown that he is indeed an expert in his own field, and the procedure of qualifying a witness as an expert is accomplished by a series of questions pertaining to his experience and qualifications in that particular field.

15.7 DEPOSITION

It is sometimes important for testimony to be introduced when it is impossible for the witness to appear personally. In such cases the testimony is recorded, under oath, with an officer of the court or the judge himself present. Attorneys from each side participate so that the resulting testimony consists of direct and cross-examination, exactly as it would be done in court. The deposition is then read in court by a clerk and entered into the trial transcript in the same manner as any other evidence. The personal touch of the witness is lacking, however, and depositions are used as little as possible.

15.8 EXPERT WITNESS

Engineers and other technical men are often requested to act as expert witnesses in a variety of court cases and hearings before official boards of inquiry. In some cases the expert is selected by the court to help explain technical matters and reach considered conclusions after hearing the facts, but in most cases he is chosen by one of the parties to a controversy and is expected to work for the party that selected him. Since his testimony may be extremely important to the outcome of a case, or may simply be used to reinforce the position of one of the parties, there are several manners in which his services may be of value to his client. Any combination of the possibilities outlined in the following paragraphs may be used as circumstances dictate.

Before commencing a suit the services of an expert may be sought in preparation of the plaintiff's petition. A study of the available facts should be made to determine whether or not there is enough evidence to warrant the suit and, hopefully, to win it. At this point the expert may be able to suggest additional facts that might be of value, or he may advise the client that he should abandon the suit. At the same time, an expert employed by the defendant may be performing the same services for him. Each party's expert should investigate the situation as completely as possible, locate and anticipate any flaws that might affect his client's position, and give him the best possible advice on whether to continue with the suit or not.

The testimony of any witness consists of answering questions asked during direct and cross-examination, and in the preparation of these questions the advice of an expert can be very useful. Such questions may be prepared by the witness himself or possibly by another expert whose knowledge is valuable but who may not perform well as a witness. The technical man and the lawyer should work together in this preparation, since the lawyer may be well versed in the technique of questioning and the ability to anticipate objections, but the expert's knowledge of the facts that must be presented should be utilized. Presentation and explanation of technical data to laymen is often very difficult and the lawyer and expert, working together, can often devise visual aids, such as models, drawings, charts, and graphs, to help in the presentation.

The presence of the prospective witness himself or a similarly trained expert during the trial itself can be very helpful to an attorney. By hearing previous witnesses and their testimony, changes in the testimony yet to be given may be suggested by the expert. The expert may also be able to find flaws in the testimony of witnesses for the other side and suggest a line of cross-examination of those witnesses that would be advantageous to his side.

An expert witness is entitled to give opinion evidence, but before such evidence is accepted it must be shown that the intended witness is an expert in the subject matter at hand. This qualifying of a witness is accomplished during his direct examination by questioning him concerning his age, education, degrees held, membership in professional and honorary societies, his work experience and reponsibilities, specialties within his profession, and any other information that might convince a jury of his high qualifications. It should be kept in mind during this questioning that the opposing lawyer may try to twist some of the information given to show that the witness in not really an expert after all, since this is one way to cast doubt on the value of his testimony. Any weakness, real or imagined, that can be anticipated as a weapon against the expert should be brought out during the qualifying questions and used as an asset before the cross-examiner is allowed to give it the appearance of a liability. The age of the witness, for example, might show that he is too young to have had much experience or so old that his training is obsolete. Scholarly writing and evidence of research can be used to show that the witness is either a person of great knowledge or is too preoccupied with his research to gain any practical experience in his field, or his experience may be either too broad or too specialized.

The conduct of the expert during both direct and cross-examination can do much to convince the jury of his credibility. He should be absolutely sure of his facts and be able to answer questions in a convincing manner. He must be able to explain technical terms and theories to the laymen of the court and jury in simple and everyday language. He should answer only what is asked and should not volunteer information. He should not guess and must be prepared to back up all his conclusions with calculations.

Very often the poise and confidence of a witness shown during direct examination when he is answering his own or otherwise carefully prepared questions may be completely shattered by the cross-examination. The opposing attorney may lead him into somewhat unfamiliar topics or tempt him into making quick mental calculations or decisions in the hope that he may make some small mistake. Any error at this time may provide an opportunity for the cross-examiner to cast doubt on all his previous correct testimony. The witness should not allow himself to become excited, hurried, or angry since the purpose of such tactics is to confuse and lead him into a damaging position. He should allow time for his own attorney to interpose objections if the questioning becomes hostile.

15.9 ARBITRATION

Small disputes between the owner and contractor are often settled by the architect or engineer, but conflicts pertaining to contract interpretation, breach of contract, liquidated damages, claims for delays and extras, and determination of pay quantities in unit-price work are serious enough to require legal action. In such cases arbitration may be used to settle the dispute rather than submit to the time-consuming expense of a lawsuit.

Provision for arbitration should be made in the contract, but unless state law stipulates that such a contract clause is binding, either party may refuse to be bound by the arbitration requirement and may elect to take the dispute to court. The general procedure for arbitration may be stated in the contract, or the contract may refer to the rules of some group such as the American Arbitration Association or the American Institute of Architects. The former group, for example, not only has a code of procedure, but will also administer the entire case or may assist only as requested.

Once a decision to arbitrate has been made, arbitrators—usually three in number—must be chosen. A common arrangement is for each side to choose one arbitrator and these two then choose the third, who may automatically become the chairman of the arbitration board. The arbitrators should be impartial and free of any conflict of interest and of course should have some knowledge of construction matters. Choosing the arbitrators from a list prepared by a disinterested outside agency often speeds up the process.

The general procedure for arbitration is similar to that of a courtroom trial but is usually much less formal. Lawyers and expert witnesses for each side may be employed, evidence is presented by witnesses for each side, and the right of cross-examination is maintained. Since the members of the board must act as both judge and jury, they must be satisfied that they have all the facts in the case. They may therefore request additional evidence and they may question the witnesses.

After all the evidence has been presented, the board, usually by a simple majority vote, makes its decision in favor of one of the parties, and awards damages or other compensation as required. No appeal of the decision to the courts is usually allowed unless fraud or some other equally serious action by the board can be proved.

The cost of the arbitration is usually shared equally by the parties to the dispute. Provision for paying the arbitrators should be established by the contract, and sometimes the amount of this compensation is set by the arbitrators themselves.

Bibliography

Qualitative, Nontechnical

ABBETT, ROBERT W., *Engineering Contracts and Specifications*. New York: John Wiley & Sons, Inc., 1954.

CLOUGH, RICHARD H., *Construction Contracting*. New York: John Wiley & Sons, Inc., 1960.

DUNHAM, CLARENCE W., and ROBERT D. YOUNG, *Contracts, Specifications, and Law for Engineers*. New York: McGraw-Hill Book Company, 1971.

McCULLOUGH, CONDE B., *The Engineer at Law*. Ames, Iowa: The Iowa State College Press, 1946.

MEAD, DANIEL W., *Contracts, Specifications and Engineering Relations*. New York: McGraw-Hill Book Company, 1956.

NORD, MELVIN, *Legal Problems in Engineering*. New York: John Wiley & Sons, Inc., 1956.

RUBEY, HARRY, and WALKER W. MILNER, *Construction and Professional Management*. New York: Macmillan Publishing Co., Inc., 1966.

RUBEY, HARRY, JOHN A. LOGAN, and WALKER W. MILNER, *The Engineer and Professional Management*. Ames, Iowa: The Iowa State University Press, 1970.

VAUGHN, RICHARD C., *Legal Aspects of Engineering*. Englewood Cliffs, N.J.: Prentice-Hall, Inc., 1962.

Quantitative, Technical

ALLIS-CHALMERS MFG. CO., *Earthmoving and Construction Data.* 1953.

AMERICAN CONCRETE INSTITUTE, *Recommended Practice for Concrete Formwork*. Detroit, Michigan, 1968.

AMERICAN PLYWOOD ASSOCIATION, *Plywood for Concrete Forming.* Tacoma, Wash., 1971.

BADZINSKI, STANLEY, *Carpentry in Commercial Construction*. Englewood Cliffs, N.J.: Prentice-Hall, Inc., 1974.

BADZINSKI, STANLEY, *Carpentry in Residential Construction*. Englewood Cliffs, N.J.: Prentice-Hall, Inc., 1972.

BENSON, BEN, *Building Contractor's and Home Builder's Handbook of Bidding, Surveying, and Estimating*. Englewood Cliffs, N.J.: Prentice-Hall, Inc., 1968.

BENSON, BEN, *Critical Path Methods in Building Construction*. Englewood Cliffs, N.J.: Prentice-Hall, Inc., 1970.

CATERPILLAR TRACTOR CO., *Fundamentals of Earthmoving*. Peoria, Ill.

CONSTRUCTION MANUFACTURERS ASSOCIATION, *Power Crane and Shovel Association Technical Bulletins 1, 2, 3, and 4.* Milwaukee, Wis., 1966.

DOUGLAS, CLARENCE J., and ELMER L. MUNGER, *Construction Management*. Englewood Cliffs, N.J.: Prentice-Hall, Inc., 1970.

HORNUNG, WILLIAM J., *Estimating Building Construction: Quantity Surveying*. Englewood Cliffs, N.J.: Prentice-Hall, Inc., 1970.

HOYLE, ROBERT J., *Wood Technology in the Design of Structures*. Missoula, Mont.: Mountain Press Publishing Company, 1973.

HUNTINGTON, WHITNEY C., *Building Construction*. New York: John Wiley & Sons, Inc., 1963.

LeTOURNEAU–WESTINGHOUSE CO., *Earthmoving, an Art and a Science*.

MEANS, ROBERT S., *Building Construction Cost Data*. Duxbury, Mass.: Robert Snow Means Co., Inc., 1973.

MOSELLE, GARY, *National Construction Estimator*. Los Angeles: Craftsman Book Company, 1971.

PEURIFOY, ROBERT L., *Construction Planning, Equipment, and Methods*. New York: McGraw-Hill Book Company, 1970.

PEURIFOY, ROBERT L., *Estimating Construction Costs*. New York: McGraw-Hill Book Company, 1958.

PEURIFOY, ROBERT L., *Formwork for Concrete Structures*. New York: McGraw-Hill Book Company, 1964.

RADCLIFFE, BYRON M., DONALD KAWAL, and RALPH J. STEPHENSON, *Critical Path Method*. Chicago: Cahners Publishing Company, Inc., 1967.

SHAFFER, L.R., L.B. RITTER, and W.L. MEYER, *The Critical Path Method*. New York: McGraw-Hill Book Company, 1965.

WYNN, A.E., and G.P. MANNING, *Formwork for Concrete Structures*. London: Concrete Publications Limited, n.d.

Index